SPACE

The History, The Mystery, The Oddities, The Curiosities

Buddy Johnston

Printed in the United States of America

Website: www.buddyswebsite.net

Dedication

To the ones who look up before they look down

to the builders with steady hands and the dreamers with stubborn hearts,

to families who kept the porch light on through long nights of launch windows and missed dinners,

to teachers who turned chalk dust into stardust,

and to every quiet soul who ever stepped outside, felt the cool dark press close, and whispered, “More.”

To the navigators who trusted numbers,

the crews who trusted each other,

and the ground teams who taught the world that excellence can be calm.

To those we lost along the way, your names are constellation lines we follow home.

And to the next generation: keep your checklist tidy, your courage humble, and your wonder loud.

The sky isn’t a ceiling. It’s a handshake. Take it.

Contents

Introduction

Step outside on a clear night. Let your eyes adjust. The sky opens like a black book with bright handwriting. Those points of light aren't decoration; they're places with rules about gravity, weather, history. Space isn't empty; it's structure, time, and story. This book is about that story and our part in it. It's not a hype parade or a trivia dump. It's a straight walk through what space is, how we learned to live with it, and why it keeps calling us forward. We'll respect the old craft of checklists, testing, calm voices, and welcome new tools that make us better, not just busier.

Dip in anywhere or go straight through. Sidebars define terms; timelines keep dates honest. When the story gets technical, we slow down and stay plain. If a line earns poetry, it gets it. If a topic needs a wrench, it gets a diagram. Maxims echo throughout: measure twice, burn once; listen to bad news; test like you fly; wonder is earned by work. No mysticism, no despair. Space is hard and fair. Good planning beats good luck. We'll hold two truths at once: boldness with humility; ambition with care. Because space quietly supports daily life of weather, GPS, crops, comms. Because exploration keeps free people sharp through curiosity with a conscience. Because Earth from orbit turns arguments into plans: borders blur; rivers, reefs, smoke, and storms do not. And because tradition isn't a museum; it's a toolkit passed to new hands.

You'll meet the famous and the essential: mathematicians who closed loops, technicians who torqued bolts, operators with cool voices on hot loops, teachers who aimed students upward, families who kept the porch light on. We honor them by showing the exact work that made big things possible. Leaving Earth, we promise to bring our best habits: count carefully, speak up, argue honestly, then commit. Treat risk with respect.

Go to learn, not posture. Widen the circle of who gets to go and who gets to decide. Keep that promise, and the sky answers with crisp data, quiet awe, and a future worth building. Turn the page. The night isn't empty. It's patient, ready, and closer than it looks.

Part I - The History

Chapter 1 - First Fires

1) Rockets, Dreams, and Accidents that Worked

Modern rocketry began with a few stubborn thinkers. Konstantin Tsiolkovsky wrote the math that proved rockets could reach space, but he warned the price would be harsh: you must carry most of your fuel just to move the rest of your fuel. Robert Goddard built small liquid-fueled rockets in farm fields and learned by failing fast and fixing faster. Hermann Oberth spread the ideas across Europe, inspiring clubs and early tests.

World War II forced rocket research to grow up. The German V-2 was a terrible weapon but a technical leap. After the war, rocket experts and documents moved to the United States and the Soviet Union. There, rockets became tools for long-range missiles. Spaceflight rode on the back of defense spending. The key idea was simple: if a rocket can throw something far, it can also throw something *up* and keep it there.

This early era set the culture of space engineering: test, measure, improve, repeat. It also taught a hard lesson that still applies: rockets don't forgive lazy math or wishful thinking.

2) Sputnik's Beep and the Shock That Shaped an Era

On October 4, 1957, the Soviet Union launched Sputnik 1, a shiny metal sphere with a radio that went "beep... beep..." as it circled Earth. The world heard those beeps and understood the message: if they can launch a satellite, they can launch heavier things. The "Space Age" was suddenly real, and it scared and inspired people at the same time.

In the United States, schools boosted science and math. Congress funded research. A new civilian space agency, NASA, was formed in 1958 to lead peaceful space exploration. Culturally, Sputnik made space dinner-table talk. Children built model rockets. Newspapers ran stories about orbits and launch pads. The sky had changed from a ceiling to a racetrack.

3) Gagarin Smiles; America Answers

On April 12, 1961, Yuri Gagarin became the first human in space. He orbited once and returned safely. His grin became a symbol of Soviet pride. The United States moved fast. On May 5, 1961, Alan Shepard made a short suborbital flight. In 1962, John Glenn completed the first American orbit.

President John F. Kennedy then set a bold goal: land a man on the Moon and return him safely to Earth before the decade ended. This wasn't just about science. It was about proving that a free society could organize, build, and succeed at a very hard task. The promise fixed the deadline. The deadline drove the budget. The budget hired the people who made the miracle.

Chapter 2 - The Moon and the Mirror

4) Apollo: Hardware, Heroism, and the Bill

The Apollo program was a stack of big, risky bets. The Saturn V rocket was taller than a 30-story building. It burned oceans of fuel in minutes. The Command Module carried the crew to lunar orbit and home again. The Lunar Module, odd and spidery, was designed only for space, not for air. It looked strange because it was perfect for its job.

Apollo cost a lot. At its peak, NASA took over 4% of the federal budget. But the money didn't vanish. It trained a generation of engineers, built new factories, and pushed computers and materials forward. Teams learned to plan, test, and think through the worst case. Those skills spread into many other industries.

In July 1969, Apollo 11 landed on the Moon. Six missions touched down. Apollo 13 failed to land but succeeded in survival, proving the system's strength. By 1972, the Moon had footprints, tire tracks, and a permanent place in human memory.

5) The Cultural Moon: How Apollo Rewired Earthly Imagination

Apollo changed the way we look at Earth. Photos like "Earthrise" and "Blue Marble" made our planet seem small and precious. Environmental movements found powerful symbols in those images.

Apollo also changed how we think about teamwork. The cool focus of Mission Control became a cultural model: calm voices, checklists, "go/no-go" decisions. We learned that heroes wear spacesuits **and** headsets.

Finally, Apollo proved that myth and measurement can live together. We still feel wonder, but now that wonder includes geology hammers, guidance computers, and careful procedures.

6) From Superpower Theater to Détente: Apollo–Soyuz

In 1975, an American Apollo and a Soviet Soyuz docked in orbit. The handshake between the crews was real and symbolic. It showed that, even during the Cold War, space could be a place for cooperation. Engineers also tested docking systems that later helped future international missions. The message was simple: you don't have to agree on Earth to share air and data in orbit.

Chapter 3 - Stations, Shuttles, Setbacks

7) Skylab and Salyut: Apartments in the Sky

Skylab, launched in 1973, was America's first real home in space. It had room for science and for learning how to live off-Earth. A damaged shield during launch forced a rescue plan that used a "parasol" to control temperature. Crews studied the Sun, Earth, and their own bodies. The Soviets ran Salyut stations, and later Mir. They focused on long stays and practical lessons: exercise daily, manage supplies, fix what breaks, and respect the clock.

Stations taught the truth about space: to learn a place, you have to *stay* there. Short visits teach you how to survive. Long visits teach you how to live.

8) The Space Shuttle: Reusable Ideal, Irreducible Risk

The Space Shuttle first flew in 1981. It was part rocket, part airplane. It promised routine flights, quick turnarounds, and lower cost. The reality was tougher. Reuse helped, but refurbishing a spaceship is never simple or cheap.

Even so, the Shuttle did amazing work. It launched and repaired satellites, especially the Hubble Space Telescope, and built major parts of the International Space Station. It also turned astronauts into construction workers, mechanics, and lab techs. The Shuttle era proved we could do complex jobs in space again and again. But it also showed that "routine" never means "easy."

9) The Price of Fire: Challenger, Columbia, and Institutional Memory

In 1986, **Challenger** broke apart after launch. In 2003, **Columbia** was lost during reentry. Both tragedies had technical causes, but they also revealed cultural problems: warnings were ignored, pressure to stay on schedule was strong, and risky conditions were normalized.

The lessons became part of spaceflight's culture: test like you fly, encourage dissent, and stop when the data looks wrong. Space is unforgiving. Procedures matter. Speaking up saves lives. These rules are now baked into training and reviews across agencies and companies.

Chapter 4 - The International Habit of Orbit

10) Mir: The Orbital Pioneer Camp

The Mir space station, run by the Soviet Union and later Russia, was active through the 1990s. It wasn't fancy, but it was tough. It grew like a treehouse, one module at a time. Crews lived and worked there for months, dealing with tight spaces, equipment failures, and the daily grind of maintenance. American astronauts trained and lived on Mir, preparing for bigger joint projects. Mir proved that long-term living in orbit is possible, but it demands patience, creativity, and constant care.

11) Building the International Space Station: Politics in Microgravity

The International Space Station (ISS) began assembly in 1998. It is a partnership among the United States, Russia, Europe, Japan, and Canada. Building it took dozens of missions, spacewalks, and careful politics. The ISS is huge, about the size of a football field, and it has been continuously occupied since 2000.

The ISS is more than a lab. It is a **promise** kept, year after year. Partners have disagreed on Earth but kept the air flowing in orbit. Crews rotate, experiments come and go, and the station stays alive through careful planning and shared responsibility. It's a flying lesson in cooperation.

12) ISS Science: Sweat, Tang, and Serious Biology

People joke about space food, but the ISS does real science. Researchers study how bones weaken, how muscles shrink, how fluids shift, and how the immune system changes in weightlessness. These studies help us prepare for deep-space missions and also teach doctors on Earth.

Other experiments look at fire and fluids in microgravity. Without gravity, flames burn in round shapes, and fluids move in surprising ways. This leads to ideas for better engines, safer spacecraft, and more efficient industrial processes.

The ISS also tests technology: life-support systems, 3D printing, and advanced robotics like the Canadarm2. Every successful test reduces risk for future missions.

Chapter 5 - New Players, New Rules

13) Europe, Japan, Canada: Quiet Giants

The European Space Agency (ESA) turns many nations' efforts into one strong program. It builds launchers, Earth-observing satellites, and bold missions like the Rosetta comet mission. Japan's JAXA is steady and creative, with sample-return missions that brought pieces of asteroids home. Canada has a smaller budget but a big reputation for robotics. The "robot arm" brand is real: strong, precise, and trusted.

These partners show that space isn't only about headlines. It's about craftsmanship and consistency. A reliable instrument or a perfect robotic grasp can matter as much as a dramatic rocket launch.

14) China, India, and the Multipolar Sky

China has moved from careful steps to confident strides: human spaceflight, a space station called Tiangong, lunar landings, and planetary probes. India's ISRO has made smart, low-cost missions that still achieve big goals, from Mars orbit to lunar exploration. The sky is now multipolar with many capable players, many goals.

This can be good. More players mean more ideas, more missions, and more reasons to keep rules clear. It also means we need to act like adults: share data when possible, avoid dangerous behavior, and talk before trouble grows.

15) Commercial Space: Rockets Meet Balance Sheets

In the last decade, commercial launch companies have cut costs by reusing rocket stages and simplifying operations. Landings that once seemed impossible are now routine. Lower prices mean more satellites, more experiments, and faster progress.

Commercial space also changed the culture. Governments are still key customers, but they are no longer the only ones. Companies sell rides to orbit, deliver cargo to the ISS, and plan private stations. Spaceports feel less like monuments and more like parts of a supply chain. The romance is still there. You can't watch a booster land and feel nothing, but now it comes with invoices and service contracts.

Chapter 6 - Culture, Commerce, and the Common Good

16) Space in the Mirror of Art, Film, and Faith

Movies and books shaped how people imagine space. Some stories are careful with science; others are not. That's fine. Stories must entertain, but teachers and agencies have to correct myths. The deeper point is older than cinema: people have long looked up and felt a mix of fear and hope. Space doesn't erase faith; it often deepens it. When you see Earth as a glowing marble against black, gratitude comes naturally, whether you say it in a lab or a church.

Art also keeps us honest. It reminds us why we went up there in the first place: not just to measure, but to *mean* something.

17) The Space Economy: Satellites, Signals, and Supply Chains

Turn off space for a day and you'll feel it. GPS runs shipping, banking, and cell towers. Earth-observation satellites guide farming, track storms, and help fight wildfires. Communication satellites keep ships online and bring internet to remote towns.

This is the "boring miracle" of space: it makes everything else on Earth work better. Insurance companies price risk more accurately. Farmers waste less water. Truck fleets save fuel. None of this is flashy, but it's where the real money and value live.

18) Law Above the Air: Treaties, Debris, and Common Sense

The Outer Space Treaty of 1967 set basic rules: space is for peaceful use, no country can claim the Moon or planets, and nations are responsible for their actors, including private companies. Later agreements handled rescue, liability, and registration.

Today's problems include space junk, traffic management, cyberattacks, and resource use. The fix is not mysterious: design satellites to die safely, share where your spacecraft are, and don't create debris on purpose. Good behavior isn't just moral; it's practical. If orbit becomes a junkyard, everyone loses.

Chapter 7 - The Road Ahead

19) Back to the Moon, Onward to Mars

Returning to the Moon is not just nostalgia. It's a training ground. We need to learn how to live off-Earth for months: power systems, habitats, precision landings near resources like water ice, and steady logistics. The Moon is close enough to practice and far enough to matter.

Mars is the long game. It offers science that could rewrite our understanding of life and climate. But it is also harsh: long travel times, thin air, bitter cold, and no quick rescue. Any Mars plan must be patient and careful. The key is to build capabilities we can reuse: landers that can scale, life-support that closes its loops, and teamwork that lasts longer than a single headline.

20) What Endures: A Tradition of Boldness

Tradition doesn't mean repeating the past. It means passing down the reasons and the discipline that made the past great. The tradition of space is courage paired with craft. The first generation proved we can leave. The second proved we can stay. The next must prove we can thrive, safely, cooperatively, and with purpose.

If we keep our humor, our checklists, and our respect for facts, the sky will stay open.

We still look up for the same reason our ancestors did: to feel small, to feel guided, and to feel called. The difference is that now we can also *go*. The Space Race showed power and pride. The Space Station showed patience and partnership. The future will test our character: can we do big, hard things together for a long time, and do them well?

I believe we can. Count it down, light the candle, and keep the faith with eyes open and numbers checked.

Chapter 8 - Fire, Math, and Stubborn Dreamers

Every big leap begins with a few stubborn people who won't shut up about an idea. For spaceflight, three names matter early:

- **Konstantin Tsiolkovsky** (Russia) wrote the math that shows how fast a rocket can get if you throw mass out the back. His famous equation says speed demands fuel, and fuel demands more fuel to push that fuel. It's a harsh bargain.
- **Robert Goddard** (United States) built small liquid-fuel rockets in empty fields. He used gyroscopes to steer and learned by blowing up a lot of hardware. Neighbors laughed until the rockets rose.
- **Hermann Oberth** (Germany) explained rocket theory, inspired clubs, and kept the idea alive across Europe.

Early rocketry was a tinker's world. You tested, you failed, you patched, you tried again. The culture that grew up around rockets learned three rules: **measure everything**, **listen to bad news**, and **never trust wishful thinking**.

1) War Rockets and the Price of Speed

World War II turned rocket dreams into brutal machines. Germany built the **V-2**, the first large rocket to touch space on the way to its target. It was technically advanced and morally awful. After the war, the Allies gathered hardware, papers, and people. Experts went to the United States and the Soviet Union. Both sides wanted long-range missiles. Spaceflight rode in the back seat of defense budgets.

Here is the hard truth: peaceful space exploration was paid for, at first, by military urgency. The same engines that could throw a warhead far could also throw a satellite up. You can dislike that fact. You should also understand it. History doesn't ask for our approval before it clocks in.

2) Sputnik's Beep: The World Wakes Up

On **October 4, 1957**, the Soviet Union launched **Sputnik 1**, a shiny ball with a simple radio. It passed overhead and said, "beep." That tiny voice was loud enough to shake governments.

Parents worried, kids got curious, and leaders reached for budgets. In 1958, the United States formed **NASA**, a civilian agency for peaceful exploration to lead an organized, scientific approach. Schools leaned into math and science. Clubs formed. Newspapers ran orbit diagrams. The Space Age had arrived, whether anyone felt ready or not.

Chapter 9 - The First Human Flights

On **April 12, 1961, Yuri Gagarin** became the first human in space. He orbited once and returned alive, smiling like a new flag. The United States followed fast. **Alan Shepard** made a 15-minute suborbital hop. **John Glenn** orbited the Earth three times in 1962.

These missions were short, but they were packed with lessons: How do you control a capsule? How do humans feel and think in weightlessness? What happens to blood, eyes, and balance? The answers shaped the checklists for everything that came next.

Chapter 10 - Building the Path to the Moon

President **John F. Kennedy** set a hard goal in 1961: land a man on the Moon and return him safely before 1970. This wasn't just science. It was strategy, pride, and public proof that a free society could do organized, exacting, dangerous work.

Meeting that goal meant three stepping-stone programs:

- **Mercury** (one astronaut): Learn to fly and survive in space.
- **Gemini** (two astronauts): Practice spacewalks, long missions, and docking, skills needed for the Moon.
- **Apollo** (three astronauts): Build the giant rocket, the Moon lander, and the navigation systems to go, land, and come home.

Training was intense. Simulators were ruthless. The culture prized calm voices, clear roles, and honest numbers. That Mission Control style, shirtsleeves, headsets, "go/no-go", became an American archetype: competence without drama.

Chapter 11 - How Apollo Worked (and Why It Was Hard)

If you want to understand Apollo, imagine stacking several miracles and asking them not to wobble.

The Saturn V Rocket

- Three stages. The first burned kerosene and oxygen to punch through thick air. The upper stages burned liquid hydrogen and oxygen for high efficiency in space.
- It stood taller than a 30-story building and made the ground shake for miles.

The Spacecraft Stack

- **Command Module (CM):** Small cone where the crew rode for launch and reentry.
- **Service Module (SM):** Tankage and engine for deep-space maneuvers.
- **Lunar Module (LM):** The bug-like lander. It had two parts: a descent stage to land and an ascent stage to lift off from the Moon and rejoin the CM.

The Plan

1. Launch to Earth orbit.
2. Fire the big engine to go to the Moon (Trans-Lunar Injection).
3. Enter lunar orbit.

4. Two astronauts land in the LM. One stays in orbit in the CM.
5. Do science fast, watch the clock, and mind the fuel.
6. Launch from the Moon, dock with the CM, transfer samples, and head home.
7. Hit Earth's atmosphere at highway-to-hell speed and splash down.

Why It Was Hard

- No GPS, no internet. Guidance was done with a small computer and careful math.
- Vacuum and dust are cruel to moving parts.
- There's no tow truck on the Moon. If something breaks, you fix it with what you have or you don't fix it at all.

The Bill

At its peak, NASA held over 4% of the U.S. federal budget. That money built factories, trained engineers, and pushed computers forward. It wasn't cheap, but it wasn't wasted. The skills spread into aviation, medicine, and industry. You can still feel Apollo in the way careful companies test and document their work.

Chapter 12 - Apollo 11-Apollo 17 Wins, Setbacks & Lessons

Apollo 11 (1969):

Neil Armstrong and Buzz Aldrin landed at the **Sea of Tranquility**. Michael Collins kept the command ship safe in orbit. The first steps were careful, almost quiet. The world watched. The message was simple: promise made, promise kept.

Apollo 12 (1969):

A precision landing near the **Surveyor 3** robot. The crew proved they could "park" on the Moon with aim, not luck.

Apollo 13 (1970):

An oxygen tank exploded on the way out. The Moon landing was canceled. The crew and controllers used the Lunar Module as a lifeboat, slung around the Moon, and came home cold, thirsty, and alive. The lesson was clearer than a bell: **systems save heroes**.

Apollo 14–17 (1971–72):

Longer stays, more science, a small rover for driving around. Geologists on Earth guided astronauts on the Moon like field instructors on a radio. The samples still teach us about how the Moon and Earth formed.

Why It Ended

By 1972, the goal was met. The public had moved on. Budgets tightened. Some planned missions were canceled. It hurts to say, but democracies don't fund wonder forever. Still, Apollo left a deep legacy: a culture of testing, a web of suppliers, and a generation trained to do the hard thing correctly the first time.

Appendix - A Useful Timeline

- **1903–1930s:** Theory and small tests by Tsiolkovsky, Goddard, and Oberth lay the groundwork for spaceflight.
- **1944–45:** V-2 rockets prove large-scale rocketry works, at a terrible cost.
- **1957:** Sputnik 1 launches; the Space Age begins in the open.
- **1961:** Gagarin orbits; Shepard and Glenn fly; Moon goal announced.
- **1969–1972:** Apollo landings; human exploration reaches a peak.
- **1973–1975:** Skylab missions and Apollo–Soyuz cooperation.
- **1970s–1990s:** Salyut and Mir develop long-duration space living.
- **1981–2011:** Space Shuttle era; Hubble repair; Challenger and Columbia losses lead to safety reforms.
- **1998–present:** ISS assembly and continuous human presence in orbit.
- **2000s–present:** Commercial launch grows; costs drop; satellite services expand.
- **2020s–beyond:** Return to the Moon; planning for Mars; more international and commercial players.

Part II - The Mystery

Chapter 13 - Surroundings Worth Knowing

Step outside at night. The air cools, and the sky opens like a black book with silver letters. Most of those letters are suns. Between them lies darkness that looks empty but is crowded with light too faint to see, fields too thin to touch, and questions that refuse to sit still. Space is silent because sound needs air, yet it hums with invisible music: gravity bending paths, radiation whispering across light-years, particles winking in and out like shy ghosts.

We call it "space," as if it were just distance, but it is also time, structure, and story. It holds what we are and what we may become. And because it withholds full answers, it invites us forward. Mystery doesn't mean "unknowable." It means "worth knowing."

1) Beginnings: A Fire Without Edges

The leading tale says the universe began about 13.8 billion years ago in a hot, dense state. In its first moments, everything that would ever be was crushed into a storm of energy and possibility. Space stretched, cooled, and made room for matter. Hydrogen and helium formed first, then stars, then heavier elements inside those stars. When massive stars died, they scattered those heavier elements into space. That stardust became new stars, new worlds, and eventually, us.

What's still mysterious is not just the *what*, but the *why*. Why was there a beginning at all? What came "before" time, if time itself had a start? Our math has no breath for a "before" at $t = 0$. Some say the word "before" simply doesn't apply. Others build careful theories of earlier cycles, bubbles, or quantum seeds. For now, the first page of the cosmic book remains smudged. We can read the paragraph, but not the handwriting.

2) Dark Matter: The Weight We Can't See

Galaxies spin too fast. Their outer stars should fling outward like water from a wild sprinkler, yet they stay bound. Something heavy is holding them together. We call that invisible weight **dark matter**. It doesn't shine or absorb light. It slips through us like wind through a screen. We know it's there because of its gravity. It makes galaxies bigger than their visible parts, sculpts clusters, and bends light from faraway objects like a glass lens.

So what is it? Ideas range from new particles to faint, cold objects to something stranger about how gravity works across large scales. The honest answer is: we are still hunting. We have traps in underground labs, telescopes in space, and brains on fire. Dark matter is a reminder that the universe can hide right in front of us and still win the staring contest.

3) Dark Energy: The Push in the Nothing

If dark matter is the hidden weight, **dark energy** is the hidden push. The universe isn't only expanding; the expansion is speeding up. That means something acts like negative pressure, pulling space apart faster and faster. Dark energy may be a property of empty space itself, or it may be a sign that our picture is missing a piece.

Whatever it is, it dominates the universe. Think about that with your morning coffee: most of the cosmos is made of *stuff we named because we don't understand it*. That's not defeat. That's honesty and a good reason to keep the lights on in our labs.

4) Black Holes: Where the Map Folds

Imagine mass so concentrated that even light cannot escape. That's a **black hole**, a cliff in spacetime itself. At its edge, called the event horizon, every path forward leads inward. Time and space trade roles. Tides stretch matter. Equations sigh.

We can't see a black hole directly, but we can watch how it tugs its neighborhood: stars whip around it; hot gas glows as it falls in; jets fire from the edges like cosmic lighthouses. At the center of most galaxies sits a supermassive black hole, millions to billions of times the mass of our Sun. Ours is called Sagittarius A*, and it keeps quiet most days, like a king who prefers whispers to speeches.

The deepest riddle is the **singularity**, the point where math breaks. Our current theories shouldn't be trusted there. We need a better one that stitches quantum rules to gravity's curve. Until then, black holes are both real objects and philosophical mirrors. They ask us: how far can you trust your tools? Where does knowledge end and humility begin?

5) Time: The Current That Changes Speed

Time feels steady: breakfast, lunch, sunset, sleep. But in the universe, time flexes. Move fast and time for you slows down relative to those who stand still. Fall deep into gravity and time runs slower for you than for someone far away. Astronauts, speeding above the atmosphere, age a touch differently than we do on the ground. The differences are tiny for people and huge for stars.

What *is* time? A dimension we pass through? A book being read one page at a time? The marching drum that keeps matter in step? We can measure time and argue about it in the same breath. If that sounds like poetry, good. The universe often speaks in couplets: one line of math, one line of wonder.

6) Light: The Messenger with Limits

Light is the fastest courier allowed, and it is also the ruler we use to measure the universe. But light has rules. It can be bent by gravity, scattered by dust, or stretched by the growth of space. When we look far away, we look back in time. The dim smudge you see in a telescope may be a galaxy as it was when dinosaurs were new on Earth, or when Earth was not yet even a promise.

Light tells truth with a delay. That delay is the price of distance. Waiting is part of science. It teaches patience, and patience teaches care.

7) Worlds: The Quiet Riot of Planets

For a long time, we knew one family of planets: our own. Now we know thousands around other stars, **exoplanets**, and more arrive in the catalogs each year. Some hug their suns so tightly they complete an orbit in days. Some wander far out, cold and slow. Some are rocky like Earth; some are giant and gassy; some are puffed like bread in an oven.

The big question leaps out: **Are any of them alive?** We look for signs, oxygen that shouldn't be there unless something keeps making it, methane that appears out of balance, heat patterns that hint at weather. We also remind ourselves that life may not look like Earth at all. The danger is to search only for ourselves under different lamps.

8) Life: A Spark That Wants Company

On Earth, life shows up wherever there is energy and chemistry, from boiling vents to silent ice. That makes many scientists hopeful. If life can thrive in so many Earthly corners, maybe it can begin wherever a world gives it a chance.

Still, there is the **Fermi Paradox**: if the galaxy is old and full of stars, and if life often appears, where is everyone? Maybe intelligent life is rare, or brief, or too far apart. Maybe the galaxy is noisy but we don't yet know the language. Maybe we are early. Maybe we are late. The paradox is a mirror: it reflects our hopes and fears as much as it reflects the sky.

9) Gravity: The Soft Hand That Shapes Everything

Gravity is the weakest force you know and the one you can't escape. It pulls oceans into tides and dust into stars. It is geometry made into a tug: mass tells space how to curve; curved space tells mass how to move. In that dance, galaxies whirl, planets glide, comets swing by for a bright hello, and the long history of structure unfolds.

We often picture gravity as a chain, but it is more like a river. Step into it, and your path changes without a sound.

10) Noise and Signal: How We Listen

Space speaks across the spectrum: radio waves, microwaves, infrared, visible, ultraviolet, X-rays, gamma rays. Each band shows different stories. Dust glows in infrared. Hot gas screams in X-rays. The afterglow of the early universe whispers in microwaves. Astronomers build different "ears," collect the notes, and make a score. The art is to separate **signal** from **noise**, the pattern that means something from the static that means "try again."

Sometimes the static wins. That is part of the charm. A clean signal without struggle is a dream; a pattern pulled from confusion is a triumph.

11) The Human Eye, the Human Limit

Our eyes are made for sunlight reflected off nearby things. They are terrible at cosmic work. So we build better eyes: telescopes on mountains, instruments in orbit, detectors buried underground, and arrays spread across deserts. We also build better minds by learning together. A lone genius is a romantic story; a careful team is how the universe actually yields its secrets.

The universe is big enough to humble us and kind enough to reward cooperation. It favors the patient, the precise, and the stubborn.

12) Risk: The Price of Looking Up Close

We send robots first because robots don't mind harsh temperatures or long silences. But sometimes we send people, because people can decide, improvise, and wonder in real time. Spaceflight is brave and fragile. Oxygen must be made or carried. Water must cycle and stay clean. Tiny errors can grow into headlines. We learn the same lesson over and over: **safety is culture** before it is equipment. Procedures save lives. So does the courage to say "stop."

Yet we keep going back because each trip increases our reach. The Moon taught us how to travel. Orbit taught us how to live away. Tomorrow's stations, ships, and bases will teach us how to stay longer and go farther with care.

13) Order and Accident: How the Universe Builds

From simple rules, complex beauty appears. Gravity collects gas into clouds; clouds collapse; stars ignite; planets form from leftovers. Impacts carve craters and tilt axes. Ice brings water; volcanoes bring air. Some worlds end in fire, some in cold, some remain half-finished like drafts a writer left on the desk.

We live on a finished draft that keeps getting edits. Drift a little closer to the Sun or a little farther away and the margins tighten. The mystery isn't just that Earth works; it's that so many steps had to land inside the safe zone for so long. Luck? Design? A mix of chance and persistence? The universe refuses to footnote.

14) The Edge That Isn't There

Ask a child, "Where does the universe end?" and you'll get answers that adults are too shy to try. Some will point to a wall. Some will say it loops around. Some will say it never ends. Cosmology leans toward "no edge" in the way a sphere has no edge: you can walk forever without falling off. Space may be finite yet unbounded, or it may be endless. The trick is that geometry can hide its nature very well when you are a tiny creature measuring a giant.

We're tiny, yes, but not helpless. We map what we can, infer what we can't, and accept that mystery is not failure. It's room to work.

15) Multiverses and Maybes

Some theories suggest our universe might be one of many, each with different settings for its physics. That idea could explain why our universe seems tuned for stars and chemistry that allows life. If there are many cosmic lotteries, we woke up in a winning ticket because only winning tickets can wake up at all.

Or maybe the tuning is an illusion; maybe life finds a way across a wide range of settings; maybe our theories need a tune-up. Multiverse talk is where science leans on philosophy for a chair and a drink. The line between testable and beautiful is a thin one. We walk it carefully.

16) The Endings We Imagine

How does it all end? Slowly, perhaps, as stars burn out and galaxies drift beyond each other's reach, leaving only black holes and the cold glow of ancient light. Maybe not so slowly, if new physics nudges the rules. We may live in a universe with a quiet exit or one that writes a twist ending. Either way, endings are only scary when you think they cancel meaning. They don't. Meaning is made in the middle, and the middle is where we are.

17) The Small and the Large: A Marriage of Scales

The universe works at every scale. At the smallest, quantum rules let particles behave like waves and probabilities. At the largest, gravity draws graceful curves across billions of light-years. The puzzle is to join the two cleanly. We don't yet have the final bridge. Call it quantum gravity, call it a new map. Until we do, we practice a kind of respectful dual citizenship of two passports, one journey.

18) Dust, Breath, and Belonging

Your bones carry calcium forged in ancient stars. Your blood carries iron from a supernova's heart. When you exhale, your breath returns carbon to the local cloud. Space is not "out there." You are part of it, and it is part of you. That's not a metaphor. That's chemistry with a sense of poetry.

Belonging matters because it beats despair. When we look up, we don't see a cold, foreign emptiness. We see our larger address.

19) Wonder and Work: How Mystery Moves Us

Mystery is not the opposite of knowledge. It is the engine of knowledge. We feel wonder, we ask questions, we test answers, we repeat. Sometimes, we are wrong. Being wrong is not a sin; it is a step. The only sin is pretending to know when you haven't measured.

Space forces good habits: clear thinking, steady nerves, shared credit, and the courage to make your work strong enough to be checked by strangers. That's not just science. That's citizenship.

20) Tradition and Tomorrow

There is a tradition to this adventure. We honor old tools but sharpen them for new tasks. We keep what works, teamwork, checklists, steady hands, and we let go of what doesn't, ego, haste, and denial. Tradition, properly kept, is not a museum. It's a memory that gives muscle to the future.

Tomorrow's mysteries will not be smaller. They will be better framed. We will search for biosignatures in exoplanet skies, listen for intelligent signals with wider nets, read galaxy clusters as if they were novels, and test the edges of physics with experiments that make today's gear look like training wheels. And we will still step outside at night and fall quiet, because quiet is part of understanding.

Coda: The Sky Inside Us

If the universe has a message, it may be this: **you are allowed to ask**. You are allowed to doubt, to test, to change your mind in the light of new numbers. You are allowed to feel small without feeling unimportant. You are allowed to love the dark for the way it makes the stars show.

Space is a mystery that invites discipline. It is a riddle with graded hints. It is a silence that wants a good question. We are stardust with questions, traveling on a warm rock around a steady sun, inside a galaxy adrift in a sea of galaxies. That is not a reason to feel lost. It is a reason to feel found.

So keep looking up. The night is not empty. It's a classroom with the lights dimmed, the chalk fresh, and the lesson still being written. We have time. We have tools. We have each other. And the bell has not rung.

Part III - The Oddities

Chapter 14 - The Quiet That Isn't There

1) Light That Takes the Long Way Home

Light is the fastest thing in the universe, but it still shows up late to everything far away. When you see a galaxy, you're meeting it as it used to be. That delay turns telescopes into time machines. Even stranger: gravity can bend light, acting like a lens. Massive clusters warp the paths of distant beams and make a single galaxy appear as arcs or duplicate ghosts. It's beautiful and rude. Imagine reading a book while the pages rearrange mid-sentence. Astronomers deal with that daily.

Oddity punchline: The straightest line in space is a curve if enough mass is nearby.

2) Dark Matter: The Heavy That Hides

Galaxies spin too fast to be held together by the stars we see. Something extra, **dark matter**, must be adding gravity. It doesn't glow, it doesn't block light, and it slips through you every second. We map where it hides by how it tugs on light and stars. Picture a ghost wearing lead boots: you can't see the ghost, but you can see the footprints.

Oddity punchline: Most of the universe's mass acts like a bouncer who won't show ID.

3) Dark Energy: The Push with No Hand

As if invisible matter weren't weird enough, the universe's expansion is **speeding up**. Something, **dark energy**, is driving space to stretch faster over time. It's not wind. It's not waves. It's more like empty space having opinions. Best guess: a property of the vacuum itself. Final answer: not yet.

Oddity punchline: Nothingness isn't neutral. It votes "grow."

4) Black Holes: Where Every Arrow Points In

A black hole is a place where gravity is so strong that even light can't escape. The edge is the **event horizon**; cross it and every path forward goes deeper. Time dilates near the edge, so an outside observer can watch you slow toward a pause that never finishes. Meanwhile, from your point of view, you fall right in. Two stories, both true, depending on who's counting seconds.

Oddity punchline: The same fall is a freeze, depending on which clock you trust.

5) Time: The River That Changes Speed

Einstein ruined "universal time" in the best way. Move fast or sit deep in gravity and your clock runs slow compared to a clock far away. Space doesn't keep one beat; it keeps a jazz rhythm. GPS satellites literally correct for this so your phone knows which driveway is yours.

Oddity punchline: Yesterday and tomorrow can disagree by a few billionths and still deliver your pizza.

6) Neutron Stars: Mountains That Are a Millimeter Tall

Crush a giant star's core, and you get a **neutron star**, a city-sized sphere with more mass than the Sun. The crust is so rigid that the tallest mountains are probably millimeters high. Yet these tiny "mountains" can release enormous energy if they crack. Some neutron stars spin hundreds of times per second, sweeping beams of radio waves like cosmic lighthouses. Others wear magnetic fields strong enough to erase your credit score from across the solar system.

Oddity punchline: The flattest place in the universe can still quake like a drum.

7) Pulsars and Cosmic Metronomes

A **pulsar** is a rotating neutron star beaming steady pulses. They're so regular that astronomers use them as galactic clocks. Here's the curveball: a pulsar can "glitch", suddenly spinning a hair faster because its superfluid interior slipped. Imagine Big Ben hiccupping once every few years, and everyone cheers because the hiccup confirms your physics.

Oddity punchline: The universe owns better clocks than we do, and they occasionally burp.

8) Magnetars: When Magnets Go Too Far

Take a neutron star and give it an absurd magnetic field. Now you have a **magnetar**. Its field is strong enough to distort atomic structure. Magnetars can blast space with **giant flares** so bright they make Earth's ionosphere jump. They are cosmic "do not touch" signs, glowing politely.

Oddity punchline: A magnet so strong it edits matter.

9) Quasars: Breakfast at the Edge of the Universe

Feed a supermassive black hole enough gas and it shines, not by the hole itself, but by the **accretion disk** heating up. The brightest of these feasts are **quasars**, visible across billions of light-years. Some fire jets that travel longer than the distance from the Sun to Pluto by thousands.

Oddity punchline: The darkest objects make the loudest light when they eat.

10) Gravitational Waves: Ripples in the Rulebook

When massive objects like black holes collide, they wrinkle spacetime. Sensitive detectors can measure those ripples, **gravitational waves**, changing the length of a four-kilometer arm by a fraction smaller than a proton. That's the scale of "you're kidding." Yet the measurement works. For the first time, we don't just **see** the universe; we **hear** it.

Oddity punchline: Space itself can ring like a bell.

11) Cosmic Background: Baby Pictures of Everything

The faint microwave glow that fills the sky is the cooled-down afterlight of the early universe. It's almost perfectly even, with tiny bumps, temperature wiggles a few parts in a hundred thousand. Those wiggles grew into galaxies. The entire cosmic web began as freckles on a newborn face.

Oddity punchline: Today's galaxies are yesterday's static.

12) The Great Silence: Fermi's Uneasy Question

If the galaxy is old, big, and full of planets, **where is everybody**? Maybe intelligent life is rare. Maybe it tends to self-destruct. Maybe it goes quiet on purpose. Maybe it's common, but spread so thin that conversations are slower than geology. We don't know, and that not-knowing is the loudest sound we hear.

Oddity punchline: The universe looks set for company; the table is still empty.

13) Rogue Planets: Worlds Without Suns

Some planets don't orbit stars. They drift alone, cold and dark, warmed only by internal heat. A few might carry thick atmospheres or subsurface oceans. Imagine a sky with no sunrise, just the glow of your own volcanoes.

Oddity punchline: Homeless worlds can still be houses.

14) Hot Jupiters and Misfit Orbits

We used to assume other systems looked like ours. Then we found **hot Jupiters**, giant planets practically kissing their stars, whipping around in a few days. Some orbits are tilted or backward. Planetary systems are not neat families; they're jazz bands with eccentric uncles.

Oddity punchline: "Normal" is a rumor our solar system started.

15) Ultra-Dense Star Cities

Globular clusters pack hundreds of thousands of stars into a tight sphere. Night there would be bright, constellations turned to crowded cities. Gravity in such a cluster choreographs slow, stellar ballets where close passes reshape orbits for eons.

Oddity punchline: Some "nights" never get dark.

16) Comets: Dirty Snowballs with Long Memories

Comets are icy archives from the solar system's childhood. Heat them and they grow tails that can stretch millions of kilometers, always pointing away from the Sun like flags whipped by a solar wind you can't feel. They sometimes carry complex molecules, inviting the old question: did comets deliver ingredients for life?

Oddity punchline: Ancient dirt writes modern chemistry in sky-script.

17) Cosmic Rays: Space's Needle Rain

High-energy particles slam into Earth from all directions. They streak through you all the time, no need to duck. Some come from supernovae, some from black hole jets, and some from mysteries we haven't solved. We detect them by the light showers they spark in the atmosphere.

Oddity punchline: The sky is always snowing, but the flakes are protons.

18) Lightning on Other Worlds

Jupiter and Saturn host lightning that makes ours look like static. Volcanoes on Io throw plumes higher than Everest. On Titan, it rains methane. On Venus, it might snow metal sulfides. Weather is a local dialect; every planet speaks it loud.

Oddity punchline: Forecasts vary: methane showers by dusk, sulfur snow by dawn.

19) Tidal Weirdness: When Gravity Sculpts Heat

Moons like **Io** and **Enceladus** are kneaded by their giant planets' gravity. The squeezing creates heat, which powers volcanoes on Io and water-rich geysers on Enceladus. No sunlight required. Where there's warm water and chemistry, biologists perk up.

Oddity punchline: You can bake a world with gravity alone.

20) Rings and Ribbons: Fine Art in Orbit

Saturn's rings look solid, but they're flocks of ice and rock guided by "shepherd" moons. The rings have gaps and waves, delicate patterns carved by gravity's invisible fingers. Other planets have rings too, but Saturn dresses best.

Oddity punchline: Sculptors in this gallery use mass, not marble.

21) The Vacuum That Isn't Empty

"Empty space" seethes with virtual particles popping in and out of existence. Most cancel out quickly, but the math hints that the vacuum has energy. That could be related to dark energy or a sign we're still missing a page. Either way, nothing is busy.

Oddity punchline: The void keeps a side hustle.

22) Antimatter: The Evil Twin That Pays Cash

Every particle has an opposite twin with opposite charge. When matter meets antimatter, they annihilate into energy. The early universe should have made equal amounts of both, but we live in a matter-dominated universe. Somewhere, a tiny preference tipped the scales. That unfair coin flip made stars, planets, and pizza possible.

Oddity punchline: The cosmos cheats slightly in our favor.

23) Quantum Whispers in a Giant Room

At small scales, particles act like waves and probabilities. At large scales, gravity is king. Strangest of all: tiny quantum ripples in the early universe may have seeded the large-scale structures we see today. The smallest rules wrote the largest patterns.

Oddity punchline: Butterflies didn't cause hurricanes. Quantum fuzz did.

24) Solar Flares and Space Weather

The Sun ties itself in magnetic knots. When a knot snaps, it spits flares and coronal mass ejections, clouds of charged particles that can shake Earth's magnetic field, paint the sky with auroras, and threaten satellites and power grids. Our calm yellow star has a dramatic streak.

Oddity punchline: The Sun smiles sweetly and occasionally throws furniture.

25) The Smell of Space (Sort Of)

Astronauts report that spacesuit gear and airlocks smell like hot metal, welding fumes, or seared steak after a spacewalk. The likely culprits are complex molecules made in high-energy reactions that cling to surfaces and then meet oxygen inside. The scent of vacuum is cooked chemistry.

Oddity punchline: Space has a "barbecue" afterglow with no grill in sight.

26) Colors You Can't See

Much of the universe glows in light our eyes can't detect: infrared from dust, ultraviolet from hot young stars, X-rays from million-degree gas. False-color images let us "see" this hidden music. The colors are chosen, yes, but the structures are real.

Oddity punchline: The prettiest space photos are honest lies that tell the truth.

27) Exoplanet Eclipses: Reading a World's Air by Starlight

When a planet crosses its star, a thin slice of starlight filters through the planet's atmosphere. That light carries a barcode, absorption lines, that tells us what gases are present. It's like diagnosing a patient from a shadow. Someday, those barcodes might spell oxygen, methane, or other signs of life out of balance.

Oddity punchline: A whisper in light can betray an ocean.

28) Galactic Collisions: Slow-Motion Grace

Galaxies collide often in cosmic time. Stars almost never hit because the spaces between them are so vast, but gravity reshapes everything, tails of stars, new bursts of star formation, and finally a peaceful merge. Think ballroom dancing at one step per million years.

Oddity punchline: Wrecks that make gardens.

29) The Axis That Isn't Up

Earth's spin axis wobbles, and the planet tilts about 23.5°. That tilt gives us seasons; the wobble shifts climate patterns across thousands of years. Other planets tilt differently. Uranus lounges at 98°, practically rolling. Its seasons are extreme: decades of daylight, decades of night.

Oddity punchline: Some worlds don't spin. They nap sideways.

30) The Edge of the Heliosphere

The **heliosphere** is the Sun's bubble, where the solar wind pushes against interstellar gas. Spacecraft like Voyager have crossed its boundary and tasted the local galactic neighborhood. The solar system, it turns out, sails in a slow cosmic breeze.

Oddity punchline: Our "shoreline" is a pressure balance drawn in plasma.

31) Gravity Assists: Slingshots with No Strings

Spacecraft steal a little momentum from planets by flying past in the right direction. It's legal cosmic shoplifting. With clever planning, you can tour several worlds on a fraction of the fuel you'd otherwise need.

Oddity punchline: The best engines include math and timing.

32) Tatooine Skies: Two Suns, Many Stories

Binary stars are common. Some planets orbit both stars; others orbit just one while the second star pirouettes nearby. Sunsets on such worlds would be complicated and excellent. Stability depends on distance and mass, cosmic choreography with strict steps.

Oddity punchline: Double the sunsets, double the headaches.

33) Snow Lines and Recipe Zones

In young disks around stars, temperature zones decide what materials can condense. Beyond certain distances, the **snow lines**, water, carbon dioxide, and other volatiles freeze, changing how planets grow. Where you start in the disk writes your future.

Oddity punchline: Planetary personality begins with freezer placement.

34) Planetary Rings That Come and Go

Rings can form from shattered moons, captured comets, or leftover dust. They can also fade as particles drift or clump into new moons. A ring is not a permanent decoration; it's an ongoing negotiation.

Oddity punchline: Jewelry with a half-life.

35) The Universe's Speed Limit Is Also Its Golden Rule

Nothing outruns light in a vacuum. That limit doesn't just slow travel, it guards cause and effect. It's why your choices can't rewrite yesterday. The speed of light is the border patrol for logic.

Oddity punchline: Physics defends the timeline with radar guns.

36) The Fine Print on "Nothing"

Vacuum experiments and theory both say emptiness has structure. Quantum fields fill space, even when they're in their lowest-energy state. That ground state can still nudge things, changing atomic levels, tugging on mirrors, or, on cosmic scales, influencing expansion. Nothing isn't a blank; it's a fabric.

Oddity punchline: The void comes pre-installed with settings.

37) Moons That Out-Moon the Planets

Some moons are more interesting than nearby planets. Europa might have a salty ocean under ice. Titan has rivers and lakes of methane. Ganymede has a magnetic field. If you're hunting for life, a moon might beat a planet without breaking a sweat.

Oddity punchline: The sidekicks are stealing the show.

38) Star Nurseries: Clouds that Give Birth

In cold molecular clouds, gravity wins over pressure and the cloud fragments into clumps that become stars. Newborn stars blast winds and light that sculpt pillars and bubbles. The place looks like a cathedral built by thunder.

Oddity punchline: Chaos writes symmetry, then vandalizes it for fun.

39) Metal Rules (Astronomer Edition)

Astronomers call any element heavier than helium a "metal" (sorry, chemists). A star's "metallicity" changes how planets form and what kinds are likely. Rich metal, more rocks. Poor metal, fewer. It's a weird vocabulary, but it predicts real worlds.

Oddity punchline: In space, oxygen is "metal." Don't tell your teacher.

40) The Universe Doesn't Do Center

People used to put Earth in the middle. Modern cosmology shrugs: **no center, no edge** in any simple sense. Expansion happens everywhere at once, the way a loaf of raisin bread expands in the oven. Pick any raisin; from its view, all the others recede. Congratulations, you're special, just like everyone else.

Oddity punchline: Everywhere is "here" to someone.

Interlude — How to Read Weirdness Well

1. **Trust measurements over moods.**
2. **Let models be maps, not the land.**
3. **Keep jokes handy.** The universe is serious; you don't have to be grim.
4. **Tradition matters.** We stand on careful methods built by people who wrote rules in sweat and, sometimes, in grief. Keep the craft. Improve the tools.

41) Space Smog: Debris and Traffic Rules

Low Earth orbit is crowded with dead satellites, paint chips, lost bolts, and retired rocket parts. A chip moving at 7.8 km/s can ruin your day.

Agencies now track debris and plan maneuvers like drivers checking mirrors. The odd part is how **orderly** the chaos can be with good cooperation.

Oddity punchline: Littering is physics with legal fees.

42) Exomoons: The Worlds That Might Be Worlds

We've found thousands of exoplanets, but exomoons are harder. Detecting a moon's subtle tug on its planet's transit timing is like hearing a mouse squeak during a thunderstorm. Hints exist; confirmations are rare. The hunt continues.

Oddity punchline: The quietest worlds may be the most common.

43) Chaos with Boundaries: Three-Body Mayhem

Two bodies orbit peacefully. Add a third, and predictions get messy. The **three-body problem** can produce dances, slingshots, and ejections. Yet out of chaos, stable pockets emerge. Navigation is half art, half calculus.

Oddity punchline: Gravity loves drama but respects choreography.

44) Deep Time: Clocks Older Than Earth

White dwarfs cool like cosmic embers. Radioactive isotopes tick inside rocks. Pulsars beat, galaxies drift. We blend these clocks to cross-check ages. The result is a universe old enough for complicated stories, but young enough to keep making new ones.

Oddity punchline: Time is a choir; no soloist can carry the song.

45) The Human Oddity: We Keep Going

Here's the strangest thing: fragile primates ride controlled explosions into orbit to study explosions they didn't cause. We're soft and small; the universe is hard and huge. Yet we launch again, wiser each time. Tradition lends backbone. Curiosity points the way. Hope does the counting.

Oddity punchline: Against odds, we show up—with checklists.

Coda — The Night That Leans Closer

Space isn't just strange; it's generous. It hides, but it also hints. It confuses, but it also teaches. The oddities aren't obstacles. They're invitations: to measure better, argue kinder, and reach farther without forgetting the ground we stand on. Keep your humor. Keep your faith in careful work. Keep your eye on the stars and your hand on the railing.

Part IV - The Curiosities

Chapter 15 - The Sky That Smiles Back

Step outside and look up. The night is a black ocean scattered with salt. It looks calm. It isn't. Stars are exploding, galaxies are colliding, and time itself is arguing over who is late. The more carefully we look, the stranger it gets, and the more sense it somehow makes. This is your field guide to cosmic oddities: plain words, sturdy facts, a little poetry when the view deserves it.

1) Gravity Bends Light: The Universe Wears Glasses

Mass curves space. Curved space bends light. That bending creates **gravitational lenses**, nature's own telescopes. A far galaxy can appear as an arc, a ring, or a string of twins. Astronomers use these lenses to see even farther, as if the universe kindly loaned us reading glasses.

Curiosity punchline: The straightest line in space is a curve if there's enough mass nearby.

2) The Heaviest Stuff Is Invisible: Dark Matter

Galaxies spin too fast to be held together by the stars we see. Something extra, **dark matter**, adds gravity without shining. We map it by how it tugs on things. It's like watching curtains billow and knowing a breeze is there, even if you can't see the air.

Curiosity punchline: Most of the cosmos is a weight you can't point to.

3) Expansion Speeds Up: Dark Energy's Quiet Push

The universe isn't just growing; it's accelerating. We blame **dark energy**, a pressure in empty space that acts like a cosmic "more, please." We can measure the push, but we don't yet know the engine.

Curiosity punchline: Nothingness is not neutral. It nudges.

4) Black Holes: Where "Forward" Means "Inward"

A **black hole** is mass folded so tight that even light can't escape. At the **event horizon**, all routes lead down. Near the edge, time stretches. To an outside observer you slow; to yourself you fall. Two clocks, two truths, one reality.

Curiosity punchline: Cross the line, and every step is down even if you think you're going straight.

5) The Universe's Baby Picture: A Whisper in Microwaves

A faint glow fills the sky in microwave light. It's the cooled **afterglow of the early universe**, almost perfectly smooth, with tiny ripples. Those ripples grew into galaxies. We can read them like a fossil of the first light.

Curiosity punchline: Today's clusters were yesterday's freckles.

6) Time Is Stretchy: Fast and Deep Means Slow

Einstein ruined the idea of one universal clock (bless him). Move fast, and your time slows. Sit in strong gravity, same deal. GPS satellites correct for this every moment so your phone knows which driveway is yours.

Curiosity punchline: Your watch and my watch are both right and slightly disagree.

7) Neutron Stars: Mountains Measured in Millimeters

Crush a star's core to the size of a city, and you get a **neutron star**. A sugar-cube of this stuff would outweigh a mountain. The surface is so rigid that the tallest "mountains" may be millimeters high—yet a crack can release more energy than a billion volcanoes.

Curiosity punchline: The flattest place in the universe can still quake like thunder.

8) Pulsars: Galactic Metronomes with the Occasional Hiccup

Some neutron stars sweep radio beams as they spin, making steady blips or **pulsars**. They keep time better than our best clocks, until they **glitch**, jumping a tiny bit faster when the superfluid interior slips.

Curiosity punchline: The universe keeps perfect time until it decides not to.

9) Magnetars: Magnets That Edit Matter

Give a neutron star an absurd magnetic field and you get a **magnetar**. Its field can twist atoms out of shape and flash the solar system from across the galaxy.

Curiosity punchline: A magnet so strong it bullies chemistry.

10) Quasars: The Darkest Things Make the Brightest Light

Feed a supermassive black hole gas and dust, and its **accretion disk** glows brighter than entire galaxies. Some launch jets longer than our solar system. They are lighthouses at the edge of time.

Curiosity punchline: When darkness eats, it shines.

11) Gravitational Waves: Space Rings Like a Bell

When black holes collide, spacetime ripples. We can **hear** these waves by measuring changes smaller than a proton over kilometers. Ridiculous yet real.

Curiosity punchline: The universe sends news without using light.

12) Fast Radio Bursts: Cosmic Morse with No Return Address

Random, bright **radio pops** crackle from distant space. Some repeat, some don't. Hypotheses multiply; certainty hides. We're listening harder.

Curiosity punchline: The sky shouts "now!" and then goes quiet.

13) Rogue Planets: Worlds Without a Sunrise

Not every planet orbits a star. Some wander alone, deep-cold and night-locked, warmed only by their own slow heat. A few might hide oceans beneath thick blankets of air or ice.

Curiosity punchline: A home can float between streetlights.

14) Hot Jupiters and Weird Orbits

We expected other systems to look like ours. Then we found **hot Jupiters**, giants skimming their stars in days, and super-Earths we don't have. Orbits can tilt, even run backward. Planetary families are jazz bands, not marching lines.

Curiosity punchline: Our solar system is normal only at home.

15) Horseshoe and Trojan Dances

Asteroids can share a planet's orbit at stable **Lagrange points** (Trojan asteroids), or do **horseshoe orbits**, slowly trading places. Gravity writes delicate choreography if you know the steps.

Curiosity punchline: Two objects, one track, no crashes most days.

16) Shepherd Moons and Sculpted Rings

Saturn's rings look smooth, but are flocks of ice and dust. Tiny **shepherd moons** herd particles, carve gaps, and craft waves. Marble was never this patient.

Curiosity punchline: The best sculptor in the solar system is gravity.

17) The Hexagon on Saturn

Winds at Saturn's north pole form a six-sided jet stream larger than Earth. Fluids can settle into stable, polygonal waves. The planet wears a geometry hat.

Curiosity punchline: Nature doodled a hexagon and kept it.

18) Blue Sunsets on Mars

Dust on Mars scatters sunlight so that the sky is butterscotch and the **sunset turns blue**, the opposite of Earth. Same physics, different dust.

Curiosity punchline: On Mars, twilight reads in cool ink.

19) Venus Spins Backwards

Venus rotates in the opposite direction to most planets and takes longer to turn once than to orbit the Sun. Its day is longer than its year, and the weather is a pressure cooker.

Curiosity punchline: A planet that prefers rewinds to play.

20) Mercury's 3:2 Beat

Mercury spins three times for every two trips around the Sun. Stand there and a single sunrise can last months.

Curiosity punchline: The clock and the calendar signed a weird treaty.

21) The Moon's Mascons and Swirls

The Moon hides dense **mass concentrations** under ancient basins that tug on passing spacecraft. It also wears pale **swirls**, curvy bright patches, likely shaped by magnetic fields deflecting solar wind.

Curiosity punchline: Even "plain" gray hides riddles.

22) Earth's Airglow, Zodiacal Light, and Noctilucent Clouds

The night sky has faint **airglow** from excited atoms, a tilted triangle of dust-scattered light called the **zodiacal light**, and high **noctilucent clouds** shining like silver fish. Space doesn't start at a hard line; it fades in.

Curiosity punchline: Darkness has a low-wattage night-light.

23) Space Has a Smell (Sort Of)

After spacewalks, astronauts say the gear smells like hot metal or seared steak. Likely cause: high-energy reactions on metal surfaces making sticky molecules that greet oxygen inside.

Curiosity punchline: Vacuum leaves a barbecue calling card.

24) Sound Without Air

Space is silent to ears, but plasma carries **pressure waves**. Instruments "hear" the Sun sing, planets hum, and shock fronts crackle. If you could plug space into a speaker, you'd get a strange, lovely choir.

Curiosity punchline: No air, still music.

25) Cosmic Rays: Needle Rain from Everywhere

High-energy particles slam into Earth all the time. They spark little **air showers** that detectors catch. Some come from supernovae, some from black hole jets, and some from mysteries still yawning.

Curiosity punchline: It's always "weather" above the weather.

26) Comets: Dirty Snowballs with Long Memories

Heat a comet and it grows a tail that always points away from the Sun. It writes bright cursive across the sky, then dives back into the cold. Some carry complex molecules, the pantry items of life.

Curiosity punchline: Ancient ice spells modern chemistry.

27) Asteroid Families and the Yarkovsky Nudge

Sunlight warms a rock; the rock re-radiates heat and gets a tiny **thrust**, the **Yarkovsky effect**, that can shift orbits over time. Family trees of asteroids tell stories of long, slow drifts.

Curiosity punchline: Sun-warmth acts like a whispering tugboat.

28) Oumuamua and Interstellar Visitors

Sometimes a body from another star system swings by. One long, thin object sped through on a weird path, reflecting light oddly and refusing to outgas like a normal comet. It left us arguing and smiling.

Curiosity punchline: The neighborhood gets surprise guests.

29) Auroras: Solar Wind in Green Silk

Charged particles from the Sun ride Earth's magnetic field into the poles, making **auroras**, curtains of green, pink, and red. A storm can make them roar to lower latitudes.

Curiosity punchline: The planet owns a private light show.

30) Space Weather Can Trip Your Lights

Big **solar storms** can disturb satellites, GPS, even power grids. We forecast space weather the way we forecast hurricanes by watching, modeling, and warning.

Curiosity punchline: The Sun is friendly until it throws furniture.

31) The Heliosphere: Our Solar Bubble

The solar wind inflates a vast **bubble** in the galaxy. At the edge, it meets interstellar gas. Spacecraft have crossed that boundary and reported the local "sea state."

Curiosity punchline: Our shoreline is drawn in plasma.

32) The Deep Space Network: Earth's Long Ears

Giant radio dishes scattered across the world talk to distant probes with whispers weaker than a nightlight on the Moon. Timing and patience are everything.

Curiosity punchline: We text robots past Pluto, and they text back.

33) The Wow! and Other Teases

Radio telescopes sometimes catch signals that look interesting and then never repeat. The most famous got labeled "**Wow!**" in a margin. We still check, still doubt, still hope.

Curiosity punchline: Space leaves voicemails we can't return.

34) Tabby's Star: The Flicker that Fooled Us All (and Helped)

A star dimmed in odd patterns. Wild ideas bloomed; careful work said dust and natural cycles. The point isn't the answer. It's the method: propose boldly, test humbly, correct publicly.

Curiosity punchline: Science blushes, then improves.

35) Biosignatures and Technosignatures

We hunt for **biosignatures**, gases out of balance, colors that hint at life, and **technosignatures**, signs of technology like strange heat patterns or radio signals. Most hints will fade under scrutiny. A few may not.

Curiosity punchline: The sky is innocent until proven chatty.

36) Dyson Spheres and Thought Experiments

Imagine a civilization building power collectors around a star. We haven't found one, but searching for odd heat is a good way to test our imagination against reality.

Curiosity punchline: Even our wild guesses need data.

37) The Coldest and Hottest Labs

Space offers **near-absolute zero** shadows on the Moon and million-degree gas near black holes. Physics gets stress-tested across absurd ranges, and the rules hold.

Curiosity punchline: One universe, many ovens and freezers.

38) The Great Cosmic Web

Galaxies aren't sprinkled at random. They line up along **filaments** and **voids** like frost veins on glass. Gravity, dark matter, and time made lace from chaos.

Curiosity punchline: Order grows where the tug is patient.

39) The Hubble Tension: Two Rulers, One Cosmos

Different methods of measuring the universe's expansion give slightly different answers. The gap is small and loud. Either we're missing a detail or a new chapter of physics is clearing its throat.

Curiosity punchline: A tiny mismatch may be a big door.

40) Deep Time and Human Hands

We compare clocks: radioactive decay in rocks, pulsar beats, white dwarf cooling, galaxy drift. Cross-checking ages is slow art. The universe is old enough for stories and young enough to surprise us.

Curiosity punchline: Truth is a choir, not a solo.

41) Microgravity's Tricks on Bodies

In orbit, fluids creep upward, bones thin, muscles sulk. Astronauts must exercise daily and plan carefully. We're learning how to live away from home without breaking.

Curiosity punchline: Your body misses weight more than you will.

42) Farming in Orbit

Plants grow oddly in microgravity. Roots guess wrong without "down," and water clings in weird blobs. Yet lettuce has been grown on the station, and seeds have made round trips. Space salads are no longer jokes.

Curiosity punchline: With the right light and care, even orbit tastes green.

43) Space Law: Own the Rock? Not So Fast.

No nation can **claim** the Moon. You can own what you extract, but the ground stays common. Launch rules, debris duties, and traffic etiquette matter more each year.

Curiosity punchline: The final frontier has fine print.

44) Calendars for Other Worlds

A day on Mars (**sol**) is about 40 minutes longer than an Earth day. Mission teams wear "Mars time" and watch sunsets creep. On the Moon, daylight lasts two Earth weeks. Timekeeping becomes a craft.

Curiosity punchline: New worlds need new watches.

45) The Human Curiosity: We Keep Showing Up

Here's the strangest part: soft primates climb into controlled explosions to study explosions they didn't cause. We are tiny, the universe is huge, and yet we keep going, over and over, wiser each time.

Curiosity punchline: Against the odds, we arrive with checklists and jokes.

Coda - Wonder with a Wrench

Curiosity is not daydreaming; it's disciplined wonder. We ask, we measure, we correct, and we try again. Tradition gives us the tools; tomorrow gives us the reason. Keep your humor. Keep your faith in careful work. Keep your eyes up and your hand on the railing.

Head Scratching Space Trivia Q & A's

1. **Q:** What's the brightest star in Earth's night sky?
   A: Sirius, in the constellation Canis Major.

2. **Q:** What's the closest star to the Sun?
   A: Proxima Centauri, part of the Alpha Centauri system.

3. **Q:** How long does sunlight take to reach Earth?
   A: About **8 minutes 20 seconds**.

4. **Q:** Which planet is the hottest on average?
   A: Venus, thanks to its intense greenhouse atmosphere.

5. **Q:** Which planet has the longest rotation period (sidereal day)?
   A: Venus, it spins once every ~243 Earth days.

6. **Q:** Which planet completes a year the fastest?
   A: Mercury, orbiting the Sun in **88 Earth days**.

7. **Q:** Which planet completes a year the slowest?
   A: Neptune, about **165 Earth years** per orbit.

8. **Q:** Which is the largest planet in the solar system?
   A: Jupiter.

9. **Q:** Which planet is least dense, so light it would float in a (ridiculously big) bathtub?
   A: Saturn.

10. **Q:** Which planet has the strongest winds?
    A: Neptune, with winds over **2,000 km/h** (1,200+ mph).

11. **Q:** What's the tallest volcano known in the solar system?
    A: Olympus Mons on **Mars,** ~22 km (13.6 mi) high.

12. **Q:** What's the huge canyon system on Mars called?
    A: Valles Marineris, thousands of kilometers long.

13. **Q:** Which world is the most volcanically active?
    A: Io, a moon of Jupiter.

14. **Q:** Which moon is the largest in the solar system?
    A: Ganymede (Jupiter), bigger than Mercury.

15. **Q:** Which moon has lakes and rivers of liquid hydrocarbons?
    A: Titan (Saturn).

16. **Q:** Which icy moon sprays water-ice geysers into space?
    A: Enceladus (Saturn).

17. **Q:** Which planet's sunsets are blue?
    A: Mars, due to fine dust scattering.

18. **Q:** Which planet rotates "backward" (retrograde) relative to most others?
    A: Venus (and **Uranus** basically rolls on its side).

19. **Q:** Which two gas giants have the most known moons?
    A: Jupiter and Saturn, each has **dozens** (over 80 each, counts keep changing).

20. **Q:** What's the only dwarf planet in the asteroid belt?
    A: Ceres.

21. **Q:** What are the two main small-body neighborhoods beyond Mars?

A: The **Asteroid Belt** (between Mars and Jupiter) and the **Kuiper Belt** (beyond Neptune).

22. **Q:** What's the (hypothetical) distant shell of comets surrounding the solar system?
A: The **Oort Cloud**.

23. **Q:** What always points away from the Sun on a comet, the dust or the tail?
A: The tail (both ion and dust tails point generally away due to solar wind and radiation pressure).

24. **Q:** What's the difference between a meteoroid, meteor, and meteorite?
A: Meteoroid = space rock in space; **meteor** = streak of light in the sky; **meteorite** = piece that **lands** on Earth.

25. **Q:** What is the Great Red Spot on Jupiter?
A: A gigantic **storm**, observed for **centuries**.

26. **Q:** What creates Earth's tides?
A: Gravity from the **Moon** (mainly) and the **Sun**.

27. **Q:** Is the Moon's "dark side" always dark?
A: No. It's just the **far side**, it gets sunlight too.

28. **Q:** What fraction of the solar system's mass is in the Sun?
A: About **99.8%**.

29. **Q:** What are the Sun's two most abundant elements?
A: Hydrogen and **helium**.

30. **Q:** How often does the International Space Station orbit Earth?
A: About **every 90 minutes** (roughly 16 orbits per day).

31. **Q:** About how fast does the ISS travel?
    A: Around **7.66 km/s (~17,500 mph)**.

32. **Q:** What was the first artificial satellite?
    A: Sputnik 1 (1957), launched by the Soviet Union.

33. **Q:** Who was the first human in space?
    A: Yuri Gagarin (1961).

34. **Q:** Who was the first American to orbit Earth?
    A: John Glenn (1962). *(First American in space: Alan Shepard, 1961, suborbital.)*

35. **Q:** Which mission achieved the first crewed Moon landing?
    A: Apollo 11 (1969).

36. **Q:** What was photographed in 2019 that we'd only theorized before?
    A: The **first image of a black hole's shadow** (M87*).

37. **Q:** What's the name of the black hole at the center of our galaxy?
    A: Sagittarius A* (Sgr A*).

38. **Q:** Do most stars live alone like the Sun?
    A: No. Many are in **binary or multiple systems**.

39. **Q:** Roughly how big is the Milky Way galaxy?
    A: About **100,000 light-years** across (order of magnitude).

40. **Q:** About how many galaxies are in the observable universe?
    A: Hundreds of billions (possibly more).

41. **Q:** What is a light-year?
    A: The **distance** light travels in one year (**~9.46 trillion km / 5.88 trillion miles**).

42. **Q:** What are gravitational waves?
    A: Ripples in spacetime made by massive accelerating objects (e.g., merging black holes).

43. **Q:** What is dark matter?
    A: Invisible mass that doesn't emit light but **adds gravity**, shaping galaxies.

44. **Q:** What is dark energy?
    A: The mysterious **cause of the accelerating expansion** of the universe.

45. **Q:** Which planet has a giant hexagon-shaped jet stream at its pole?
    A: Saturn (north pole).

46. **Q:** Which moon likely has a salty ocean beneath its ice and is a prime target in the life hunt?
    A: Europa (Jupiter).

47. **Q:** What are pulsars?
    A: Rapidly rotating **neutron stars** that beam regular pulses of radiation.

48. **Q:** What's the farthest human-made object from Earth?
    A: Voyager 1, now in **interstellar space**.

49. **Q:** What famous item did the Voyagers carry as a cultural time capsule?
    A: The **Golden Record**, sounds and images from Earth.

50. **Q:** What's currently the fastest human-made object?
    A: NASA's Parker Solar Probe (holding the speed record as it dives near the Sun).

51. **Q:** What's the Milky Way's general shape?
    A: A **barred spiral** galaxy.

52. **Q:** What's the name of our local galaxy group?
    A: The **Local Group**, led by the Milky Way and **Andromeda**.

53. **Q:** What galaxy is on a slow collision course with the Milky Way?
    A: Andromeda (M31).

54. **Q:** What are most stars made of?
    A: Hydrogen and **helium**.

55. **Q:** What do stars fuse in their cores during the longest part of their lives?
    A: Hydrogen into helium.

56. **Q:** What determines a star's color?
    A: Its **surface temperature** (hotter = bluer; cooler = redder).

57. **Q:** What is a red giant?
    A: An **aged star** that has expanded after using up core hydrogen.

58. **Q:** What stellar event creates most elements heavier than iron?
    A: Supernovae and **neutron star mergers**.

59. **Q:** What is a white dwarf?
    A: The **hot, dense core remnant** of a Sun-like star.

60. **Q:** What keeps a white dwarf from collapsing further?
    A: Electron degeneracy pressure.

61. **Q:** What is a planetary nebula?
    A: A **shell of gas** ejected by a dying Sun-like star.

62. **Q:** What is a supernova remnant?
    A: The **expanding debris** left after a star explodes.

63. **Q:** What's the Kuiper Belt's most famous dwarf planet (besides Pluto)?
    A: Eris (also: Haumea, Makemake).

64. **Q:** What are Trojan asteroids?
    A: Objects sharing a planet's orbit at **Lagrange points**.

65. **Q:** What is the main ingredient of Jupiter and Saturn?
    A: Hydrogen (with helium).

66. **Q:** What is the main ingredient of Uranus and Neptune?
    A: Ices (water, ammonia, methane) plus hydrogen/helium, hence "**ice giants**."

67. **Q:** Why is Mars red?
    A: Iron oxide (rust) dust covering the surface.

68. **Q:** What causes phases of the Moon?
    A: Our **viewing angle** of the sunlit half as the Moon orbits Earth.

69. **Q:** What's a lunar eclipse?
A: The **Moon** passes through **Earth's shadow**.

70. **Q:** What's a solar eclipse?
A: The **Moon** blocks the **Sun** from Earth's view.

71. **Q:** Why don't we get eclipses every month?
A: The Moon's orbit is **tilted** relative to Earth's orbit.

72. **Q:** What does "albedo" mean?
A: An object's **reflectivity**.

73. **Q:** Which planet has the highest albedo (very reflective clouds)?
A: Venus.

74. **Q:** What's the name of the boundary where Earth's atmosphere meets space by common convention?
A: The **Kármán line** (about 100 km up).

75. **Q:** What protects Earth from most solar wind particles?
A: The **magnetosphere**.

76. **Q:** What causes auroras?
A: Solar particles hitting **upper-atmosphere gases** along magnetic field lines.

77. **Q:** What's the approximate tilt of Earth's axis?
A: About **23.5°**, giving us **seasons**.

78. **Q:** What is an exoplanet?
A: A **planet orbiting another star**.

79. **Q:** What's a common method to discover exoplanets?
A: The **transit method**, measuring tiny drops in starlight.

80. **Q:** What's the "habitable zone"?

    A: The **distance range** where liquid water can exist on a world's surface.

81. **Q:** What is a light curve?

    A: A **graph of brightness** versus time, used to spot transits and variable stars.

82. **Q:** What is redshift?

    A: Light **stretched to longer wavelengths** as objects move away (or space expands).

83. **Q:** What is blueshift?

    A: Light **compressed to shorter wavelengths** as objects approach.

84. **Q:** What does the "cosmological principle" say?

    A: On large scales, the universe is **homogeneous and isotropic**.

85. **Q:** What is the "observable universe"?

    A: The **region we can see** because light has had time to reach us since the Big Bang.

86. **Q:** What sets the ultimate speed limit in the universe?

    A: The **speed of light in vacuum**.

87. **Q:** What is a barycenter?

    A: The **common center of mass** two bodies orbit.

88. **Q:** Does the Moon orbit Earth's center?

    A: Both Earth and Moon orbit their **barycenter** (inside Earth but offset from its center).

89. **Q:** What makes a day slightly longer over geologic time?
A: Tidal braking, the Moon's pull slows Earth's rotation.

90. **Q:** What is a synodic month?
A: New Moon to new Moon cycle (~29.5 days).

91. **Q:** What is planetary migration?
A: Planets **shifting orbits** after forming, due to disk interactions or gravity.

92. **Q:** What's special about Titan's atmosphere?
A: It's **thick** and **nitrogen-rich**, denser than Earth's.

93. **Q:** What gives Uranus and Neptune their blue color?
A: Methane absorbing red light.

94. **Q:** What is a ring system mostly made of?
A: Ice and rock particles (sizes from dust to boulders).

95. **Q:** What is space debris?
A: Human-made junk in orbit: defunct satellites, fragments, flakes.

96. **Q:** Why do satellites fall eventually from low orbits?
A: Atmospheric drag slowly reduces altitude.

97. **Q:** What is a gravity assist?
A: Using a **planet's motion and gravity** to change a spacecraft's speed and direction.

98. **Q:** Why are space suits white for spacewalks?
A: Reflect heat and improve **visibility**.

99. **Q:** What do astronauts do to avoid "the bends" before a spacewalk?

A: Pre-breathe oxygen to reduce nitrogen in blood.

100. **Q:** What is microgravity?

A: Free-fall conditions that feel like weightlessness.

101. **Q:** What's the most common type of star in the Milky Way?

A: Red dwarfs (small, cool, long-lived).

102. **Q:** What is a circumbinary planet?

A: A planet that **orbits two stars** at once.

103. **Q:** What is a comet's ion tail made of?

A: Charged gas pushed by the **solar wind** (often straight, bluish).

104. **Q:** What is a comet's dust tail made of?

A: Dust particles pushed by **sunlight pressure** (broader, curved).

105. **Q:** What is the Roche limit?

A: The distance within which **tidal forces** can tear a body apart.

106. **Q:** What happens if a moon drifts inside a planet's Roche limit?

A: It can **break up** and form **rings**.

107. **Q:** What is a tidally locked world?

A: One that **always shows the same face** to its partner (like our Moon).

108. **Q:** What is a brown dwarf?

A: An object **too big to be a planet**, too small to sustain long-term hydrogen fusion.

109. **Q:** What is metallic hydrogen suspected to explain?

A: Jupiter's strong **magnetic field** and interior structure.

110. **Q:** What is the heliopause?

A: The boundary where the **solar wind** is stopped by the **interstellar medium**.

111. **Q:** What is an Einstein ring?

A: A **perfect circle** of lensed light when source, lens, and observer align.

112. **Q:** What is a Type Ia supernova used for in cosmology?

A: As a **"standard candle"** to gauge cosmic distances.

113. **Q:** What is baryonic matter?

A: Normal matter made of protons, neutrons, and electrons.

114. **Q:** What is an active galactic nucleus (AGN)?

A: A **bright core** powered by a **feeding supermassive black hole**.

115. **Q:** What is an ultra-diffuse galaxy?

A: A galaxy with **very low surface brightness** but sometimes **Milky Way-sized**.

116. **Q:** Why does Pluto have a heart-shaped region?

A: Sputnik Planitia, a **nitrogen-ice basin** that affects Pluto's winds and spin.

117. **Q:** What is the Yarkovsky effect?

A: A **tiny thrust** from an object's heat re-radiation, slowly changing its orbit.

118. **Q:** What does "perihelion" mean?

A: Closest point to the **Sun** in an object's orbit (aphelion = farthest).

119. **Q:** What are planetary "rings" around small bodies like Chariklo?

A: Narrow ring arcs discovered around minor bodies, rings aren't just for giants.

120. **Q:** What's the big picture lesson of space?

A: The universe is **huge, lawful, and surprising** and careful measurement beats guesswork every time.

121. **Q:** What do the letters O, B, A, F, G, K, M classify?

A: Stellar spectral types from hottest (O) to coolest (M).

122. **Q:** What type of star is the Sun by spectrum?

A: G-type main-sequence star (G2V).

123. **Q:** What is the Hertzsprung-Russell diagram?

A: A plot of **stellar brightness vs. temperature** that maps stellar life cycles.

124. **Q:** What is stellar parallax used for?

A: Measuring distances to nearby stars by apparent shift as Earth orbits.

125. **Q:** What is one astronomical unit (AU)?

A: The **average Earth–Sun distance** (~150 million km).

126. **Q:** What is a parsec?
A: 3.26 light-years, the distance at which 1 AU subtends 1 arcsecond.

127. **Q:** What are Cepheid variables famous for?
A: Their **period–luminosity** relation used to measure cosmic distances.

128. **Q:** What process powers main-sequence stars?
A: Hydrogen fusion in their cores.

129. **Q:** What happens when a Sun-like star runs out of core hydrogen?
A: It **expands into a red giant**.

130. **Q:** What is a supernova?
A: A **stellar explosion** that briefly outshines a galaxy and forges heavy elements.

131. **Q:** What's left after a massive star explodes?
A: A **neutron star or black hole**, plus a remnant nebula.

132. **Q:** What is a protoplanetary disk?
A: A **disk of gas and dust** around a young star where planets form.

133. **Q:** What's the snow line in a disk?
A: The **distance** beyond which volatiles like water can **freeze**.

134. **Q:** What is planetary differentiation?
A: Separation of materials by density into **core, mantle, crust**.

135. **Q:** Why is Earth geologically active?
A: Internal heat from formation and radioactivity drives **plate tectonics**.

136. **Q:** What are the Van Allen belts?
A: Doughnut-shaped zones of **trapped charged particles** around Earth.

137. **Q:** What causes Earth's magnetic field?
A: Convection of molten iron in the outer core (geodynamo).

138. **Q:** What is an impact crater's central peak?
A: A **rebound mound** formed after the impact compresses and relaxes.

139. **Q:** Which planet has the largest canyon system?
A: Mars - Valles Marineris.

140. **Q:** What is regolith?
A: Loose, **fragmented surface material** covering solid rock (e.g., lunar soil).

141. **Q:** Why does the Moon have more craters than Earth?
A: No atmosphere, weather, or plate tectonics to erase them.

142. **Q:** What is tidal locking?
A: A body **rotates once per orbit**, always showing the same face (the Moon).

143. **Q:** What is escape velocity?
A: The **speed needed** to break free from a body's gravity without more thrust.

144. **Q:** What is a transit?

A: A planet **crossing its star's face**, dimming the star slightly.

145. **Q:** What is radial-velocity detection?

A: Finding planets via **stellar wobble** from gravitational tug.

146. **Q:** What is a habitable zone?

A: The **distance range** where surface liquid water **could** exist.

147. **Q:** What is a biosignature?

A: A **chemical or physical sign** that may indicate life (e.g., certain gas mixes).

148. **Q:** What is a technosignature?

A: A sign of **technology**, like unusual radio emissions or waste heat.

149. **Q:** What is interstellar medium (ISM)?

A: Gas and dust between stars.

150. **Q:** What are molecular clouds?

A: Dense, cold ISM regions where **stars form**.

151. **Q:** What does ALMA observe best?

A: Millimeter/submillimeter waves, great for **cold dust and gas**.

152. **Q:** What does the JWST specialize in?

A: Infrared astronomy, seeing **dust-shrouded** objects and distant galaxies.

153. **Q:** What kind of light does Chandra study?

A: X-rays from hot cosmic regions.

154. **Q:** What did Hubble famously fix?
A: Its **initial mirror flaw**, via **servicing missions**.

155. **Q:** What is an active galactic nucleus (AGN) powered by?
A: A **feeding supermassive black hole**.

156. **Q:** What are jets from black holes made of?
A: Highly energetic plasma ejected along magnetic fields.

157. **Q:** What is gravitational lensing useful for?
A: Mapping **dark matter** and magnifying distant galaxies.

158. **Q:** What are baryon acoustic oscillations?
A: Imprints of early-universe sound waves used as a **cosmic ruler**.

159. **Q:** What does "homogeneous and isotropic" mean in cosmology?
A: On large scales, matter is **evenly spread** and **looks the same** in all directions.

160. **Q:** What is the cosmic web?
A: A **filament–void** structure of galaxies and dark matter across the universe.

161. **Q:** What defines the observable universe?
A: The region from which **light has reached us** since the Big Bang.

162. **Q:** What is cosmic redshift primarily from at large scales?
A: Expansion of space stretching light.

163. **Q:** What is the Sun's approximate age?
A: About **4.6 billion years**.

164. **Q:** What's the Sun's expected total main-sequence lifetime?
A: Roughly **10 billion years**.

165. **Q:** What is the solar wind?
A: A stream of **charged particles** flowing outward from the Sun.

166. **Q:** What are coronal mass ejections (CMEs)?
A: Large eruptions of solar plasma and magnetic field.

167. **Q:** What are sunspots?
A: Cooler, darker magnetic regions on the Sun's surface.

168. **Q:** What cycle do sunspots roughly follow?
A: About **11 years**.

169. **Q:** What is a prominence?
A: A loop of **cooler, dense plasma** suspended in the solar corona.

170. **Q:** What is heliophysics?
A: The study of the **Sun and its influence** on the solar system.

171. **Q:** What is a mission "flyby"?
A: A **close pass** to gather data without orbiting or landing.

172. **Q:** What is aerobraking?
A: Using **atmospheric drag** to slow a spacecraft.

173. **Q:** What is a reaction wheel used for?
A: Pointing and stabilizing spacecraft without using fuel.

174. **Q:** What are RCS thrusters?
A: Small **reaction control** thrusters for attitude and fine maneuvers.

175. **Q:** What is a Lagrange point?
A: A spot where **gravitational forces balance**, allowing stable or semi-stable orbits.

176. **Q:** Which Lagrange point hosts JWST?
A: Sun–Earth L2.

177. **Q:** What is planetary protection?
A: Policies to **avoid contaminating** other worlds (and Earth) with microbes.

178. **Q:** What is a clean room for spacecraft?
A: A **controlled environment** to minimize contamination.

179. **Q:** What was the first soft landing on the Moon?
A: Soviet Luna 9 (1966).

180. **Q:** What mission first orbited Jupiter and studied its moons up close in the 1990s?
A: Galileo.

181. **Q:** What mission orbited Saturn and dropped a probe on Titan?
A: Cassini–Huygens.

182. **Q:** What mission flew by Pluto in 2015?
A: New Horizons.

183. **Q:** What mission returned asteroid samples to Earth from Bennu?

A: OSIRIS-REx.

184. **Q:** What mission orbited and touched down on asteroid Ryugu?

A: Hayabusa2 (JAXA).

185. **Q:** What mission landed on a comet's surface?

A: Rosetta (ESA) with **Philae.**

186. **Q:** What is the first image of a black hole's shadow called?

A: Image of **M87*** by the **Event Horizon Telescope.**

187. **Q:** What is Sagittarius A*?

A: The **supermassive black hole** at the Milky Way's center.

188. **Q:** What are spiral arms made of?

A: Stars, gas, and dust in regions of active star formation.

189. **Q:** What is an elliptical galaxy like?

A: Spheroidal, with older stars and **less gas.**

190. **Q:** What is a dwarf galaxy?

A: A **small galaxy** with far fewer stars than the Milky Way.

191. **Q:** What's the likely fate of the Milky Way and Andromeda?

A: A **merger** into a larger galaxy in billions of years.

192. **Q:** What is an ultra-luminous X-ray source (ULX)?

A: An **extremely bright X-ray** object, often a black hole or neutron star accretor.

193. **Q:** What is a kilonova?

A: A bright event from **neutron star mergers**, creating heavy elements.

194. **Q:** What are gravitational wave detectors like LIGO measuring?

A: Tiny spacetime stretches from massive mergers.

195. **Q:** What is a light curve dip outside transits often caused by?

A: Starspots or **stellar variability**.

196. **Q:** What is limb darkening?

A: A star's **edge looks dimmer** than its center due to viewing angle through its atmosphere.

197. **Q:** Why does the Moon have maria (dark plains)?

A: Ancient lava flows filled large impact basins.

198. **Q:** What is basalt on the Moon?

A: Volcanic rock forming much of the maria.

199. **Q:** What is the lunar highlands made of mainly?

A: Anorthosite, a light-colored rock rich in plagioclase.

200. **Q:** What is synchronous orbit for satellites?

A: An orbit with **period equal to the body's rotation**, e.g., geostationary around Earth.

201. **Q:** What is a geostationary satellite's altitude?

A: About **35,786 km** above the equator.

202. **Q:** What is a sun-synchronous orbit used for?

A: Imaging with consistent lighting conditions.

203. **Q:** What are CubeSats?

A: Small modular satellites built from 10-cm cubes.

204. **Q:** What is a space telescope's main advantage over ground ones?

A: No atmospheric distortion or absorption.

205. **Q:** What is adaptive optics?

A: Ground-based tech that **corrects for atmospheric turbulence** in real time.

206. **Q:** What is the zodiacal light made of?

A: Sunlight scattered by **interplanetary dust**.

207. **Q:** What is a bolide?

A: An **exceptionally bright** meteor, often with explosions.

208. **Q:** What is a fireball?

A: A **very bright meteor**, brighter than Venus.

209. **Q:** What is the Taurid or Perseid?

A: A **meteor shower** from comet debris.

210. **Q:** What is a carbonaceous chondrite?

A: A **primitive meteorite** rich in carbon and organic compounds.

211. **Q:** What is space weathering?

A: Surface changes on airless bodies from micrometeorites and radiation.

212. **Q:** Why does Mercury have extreme temperatures?

A: Thin exosphere and **slow rotation**, poor heat distribution.

213. **Q:** What is Venus's atmosphere mostly?

A: Carbon dioxide, with high pressure and clouds of sulfuric acid.

214. **Q:** What makes Mars's sky butterscotch?

A: Fine dust scattering sunlight.

215. **Q:** What gives Uranus and Neptune their blue color?

A: Methane absorbing red light (plus hazes).

216. **Q:** What is differential rotation on gas giants?

A: Different **latitudes rotate at different speeds**.

217. **Q:** What is Io's volcanism powered by?

A: Tidal heating from Jupiter's gravity.

218. **Q:** What powers Enceladus's geysers?

A: Tidal heating and subsurface **liquid water**.

219. **Q:** What makes Titan's lakes and rivers?

A: Liquid methane and ethane.

220. **Q:** What is Europa's surface like?

A: Icy, crisscrossed with **fractures**, likely covering a **subsurface ocean**.

221. **Q:** What is a shepherd moon?

A: A small moon that **gravitationally shapes** planetary rings.

222. **Q:** What is ring "spoke" activity on Saturn?

A: Dust features lifted by **electromagnetic** effects.

223. **Q:** What is a contact binary asteroid?

A: Two lobes **gently merged**, looking like a "rubber duck."

224. **Q:** What is planetary albedo important for?
A: Temperature balance and climate.

225. **Q:** What is the greenhouse effect?
A: Atmosphere **traps heat** by absorbing infrared radiation.

226. **Q:** What's special about Mercury's magnetic field?
A: It's **weak** but **global**, likely from a **partly molten core**.

227. **Q:** What is space debris also called?
A: Orbital debris or **space junk**.

228. **Q:** Why is orbital inclination important?
A: It sets **ground track** and which latitudes a satellite can see.

229. **Q:** What is a Hohmann transfer?
A: An **efficient two-burn** orbital transfer between circular orbits.

230. **Q:** What is delta-v?
A: The **change in velocity** a spacecraft needs for maneuvers.

231. **Q:** What is specific impulse (Isp)?
A: A measure of **rocket efficiency** (thrust per propellant flow).

232. **Q:** What propels ion engines?
A: Ionized propellant accelerated by **electric fields**, very efficient, low thrust.

233. **Q:** What is ballistic coefficient?
A: A body's **mass-to-drag** ratio, affects how it slows in an atmosphere.

234. **Q:** What is a skyhook or space elevator concept based on?
A: A **tether** anchored to Earth or orbit using **centrifugal force**, still theoretical.

235. **Q:** What is a cryovolcano?
A: A **ice-volcano** erupting water/ammonia instead of lava.

236. **Q:** What is a tholin?
A: Complex **organic gunk** formed by radiation in ices and atmospheres.

237. **Q:** What is the helioseismology field about?
A: Studying **waves in the Sun** to probe its interior.

238. **Q:** What is astroseismology?
A: Using **stellar oscillations** to measure star interiors.

239. **Q:** What is the faint young Sun paradox?
A: Early Sun was **dimmer**, yet Earth had **liquid water**, implies strong greenhouse effects.

240. **Q:** What is a gamma-ray burst (GRB)?
A: An intense **flash of gamma rays** from distant cataclysms.

241. **Q:** What is microlensing used for?
A: Detecting **dark objects** (and exoplanets) via **temporary brightening**.

242. **Q:** What is an Einstein cross?
A: A quasar split into **four images** by gravitational lensing.

243. **Q:** What is a bow shock around a star?
A: A **wave** where stellar wind meets interstellar medium.

244. **Q:** What is a proplyd?

A: A **protoplanetary disk** being ionized by nearby massive stars.

245. **Q:** What is solar differential rotation?

A: The Sun's **equator rotates faster** than its poles.

246. **Q:** What are spicules on the Sun?

A: Jet-like eruptions from the chromosphere.

247. **Q:** What is coronal heating problem?

A: The corona is **hotter** than the surface, an active research puzzle.

248. **Q:** What is interstellar reddening?

A: Dust **scattering blue light**, making distant objects look **redder**.

249. **Q:** What is a reflection nebula vs. emission nebula?

A: Reflection scatters starlight; **emission** glows from **ionized gas**.

250. **Q:** What is a planetary transit timing variation (TTV)?

A: Timing shifts in transits caused by **other planets' gravity**.

251. **Q:** What is an occultation?

A: One body **passing in front of another**, briefly blocking its light.

252. **Q:** What is the opposition of a planet?

A: When the planet is **opposite the Sun** in our sky, closest and brightest.

253. **Q:** What is inferior vs. superior conjunction?

A: Inferior: inner planet between Earth and Sun. **Superior:** planet behind the Sun.

254. **Q:** What is a synodic vs. sidereal period?

A: Synodic: relative to Sun/Earth alignment; **sidereal:** relative to the stars.

255. **Q:** What is the cosmic microwave background's temperature today?

A: About **2.7 K** above absolute zero.

256. **Q:** What is cosmic variance?

A: Limits to precision from **having only one universe** to observe.

257. **Q:** What is an ultra-cool dwarf?

A: A **very low-temperature** star or brown dwarf with late spectral types.

258. **Q:** What is a tidal disruption event?

A: A star **torn apart** by a black hole's gravity.

259. **Q:** What is an accretion disk?

A: Rotating material spiraling into a massive object, heating up and glowing.

260. **Q:** What is the ISRU idea for space exploration?

A: In-situ resource utilization, using local materials (e.g., lunar ice).

261. **Q:** What is a regolith "cold trap"?

A: Permanently shadowed regions that **store volatile ices**.

262. **Q:** What is cosmic ray spallation?

A: High-energy particles **break nuclei**, creating **light elements** (like Li, Be, B).

263. **Q:** What is astrobiology?

A: The study of **life in the universe**, its origins, evolution, and distribution.

264. **Q:** What is panspermia?

A: The idea that **life's building blocks** (or microbes) can **travel between worlds**.

265. **Q:** What is a heliocentric model?

A: A model with the **Sun at the center** of the solar system.

266. **Q:** What is retrograde motion (apparent)?

A: A planet's **temporary backward track** in the sky due to relative orbits.

267. **Q:** What are Kepler's three laws about?

A: Planetary **orbits**: ellipses, equal areas in equal times, and period–distance relation.

268. **Q:** What is the Roche lobe in binaries?

A: The region around a star where **its gravity dominates** in a binary system.

269. **Q:** What is a contact binary star?

A: Two stars **sharing an envelope**.

270. **Q:** What is a blue straggler?

A: A **young-looking star** in an old cluster, likely rejuvenated by **mass transfer/merger**.

271. **Q:** What is a globular cluster?
A: A **dense, spherical** star cluster orbiting a galaxy's halo.

272. **Q:** What is an open cluster?
A: A **looser** group of young stars in a galaxy's disk.

273. **Q:** What is the Great Attractor?
A: A region of **enhanced gravity** influencing local galaxy motions.

274. **Q:** What is a cosmic void?
A: A **large, low-density** region with few galaxies.

275. **Q:** What is a tidal tail in galaxies?
A: **Streamers of stars and gas** pulled out by gravitational encounters.

276. **Q:** What is ram-pressure stripping?
A: Gas **removed from galaxies** as they move through **hot cluster gas**.

277. **Q:** What is a starburst galaxy?
A: A galaxy with **very high star-formation rate**.

278. **Q:** What is metallicity in astronomy?
A: The **fraction of elements heavier than helium** in a star.

279. **Q:** Why does metallicity matter for planets?
A: Higher metallicity **correlates with more planets**, especially giants.

280. **Q:** What is an isochrone in stellar work?
A: A curve of **equal age** on an H-R diagram for a population.

281. **Q:** What is a white dwarf's typical size?

A: About **Earth-sized** but extremely dense.

282. **Q:** What is the Chandrasekhar limit?

A: Max white-dwarf mass ~**1.4 solar masses**.

283. **Q:** What is a nova?

A: A **sudden brightening** from thermonuclear runaway on a white dwarf's surface.

284. **Q:** What is a Type Ia supernova progenitor?

A: A white dwarf **reaching critical mass** (via accretion/merger).

285. **Q:** What is a magnetosphere used for by planets?

A: Deflecting solar wind, protecting atmosphere and surface.

286. **Q:** What is planetary synergy for life searches?

A: Combining **orbital data, spectra, and models** to assess habitability.

287. **Q:** What is a phase curve (exoplanets)?

A: Brightness change as a planet shows **day/night sides**.

288. **Q:** What is obliquity?

A: A planet's **axial tilt** relative to its orbital plane.

289. **Q:** What is precession?

A: The **slow wobble** of a planet's spin axis.

290. **Q:** What is nutation?

A: Smaller **oscillations** superimposed on precession.

291. **Q:** What is an opposition surge?

A: A sharp **brightening** of airless bodies when Sun-observer-target align.

292. **Q:** What is the terminator on a planet?

A: The **day-night boundary**.

293. **Q:** What is space's main heat transfer mode?

A: Radiation (no air for convection/conduction between bodies).

294. **Q:** Why do spacecraft need radiators?

A: To **dump excess heat** via radiation.

295. **Q:** What is a Mullins-Sekerka instability?

A: A crystal growth instability, relevant in **microgravity materials science**.

296. **Q:** What is cosmic microwave background polarization used for?

A: Probing **early-universe physics** and **structure growth**.

297. **Q:** What is reionization?

A: Era when early stars/galaxies **ionized** the intergalactic medium again.

298. **Q:** What is a Lyman-alpha forest?

A: A series of **absorption lines** from intergalactic hydrogen in quasar spectra.

299. **Q:** What is an exosphere?

A: The **uppermost tenuous layer** of an atmosphere where particles escape.

300. **Q:** What is hydrostatic equilibrium in stars?

A: Balance of **inward gravity** and **outward pressure**.

301. **Q:** What is a coronagraph?

A: An instrument that **blocks starlight** to see faint nearby objects.

302. **Q:** What is direct imaging of exoplanets?

A: Photographing planets by suppressing the host star's glare.

303. **Q:** What is a debris disk?

A: A ring of **dust and small bodies** around a star, planet building leftovers.

304. **Q:** What is a hot Jupiter likely evidence of?

A: Planetary migration after formation.

305. **Q:** What is a mean-motion resonance?

A: Orbital periods in a **simple ratio** (e.g., 2:1), enhancing stability or exchange.

306. **Q:** What is an oblate spheroid?

A: A **flattened sphere** from rotation (e.g., **Jupiter**).

307. **Q:** What is stellar wind stripping?

A: Atmosphere loss due to a star's particle flux.

308. **Q:** What protects Mars less than Earth from stripping?

A: No global magnetic field and **thin atmosphere**.

309. **Q:** What is Viking's big Mars result about life?

A: No confirmed life, but important **chemistry** and sterile protocols.

310. **Q:** What did Curiosity confirm about ancient Mars?
A: It once had **environments that could support microbial life**.

311. **Q:** What are Martian "blueberries"?
A: Hematite spherules formed by past water activity.

312. **Q:** What is a dust devil on Mars?
A: A **spinning column** lifting dust, common in thin air.

313. **Q:** What is seasonal CO_2 frost on Mars?
A: Dry ice forming and sublimating at the poles.

314. **Q:** What is a cryosphere?
A: The **frozen water** part of a planet's surface/subsurface.

315. **Q:** What is aerocapture?
A: A **single pass** through atmosphere to enter orbit, riskier than aerobraking.

316. **Q:** What is a ballistic trajectory?
A: A path with **no propulsion** after initial thrust (gravity does the rest).

317. **Q:** What is a Lambert solution used for?
A: Orbital transfer calculations between two points in time.

318. **Q:** What is a gravity gradient boom?
A: A **mass on a tether** for passive spacecraft stabilization.

319. **Q:** What is star tracker equipment for?
A: Determining attitude by recognizing star patterns.

320. **Q:** What is a micrometeoroid?

A: A **tiny grain** that can still damage spacecraft at high speed.

321. **Q:** What is a Whipple shield?

A: Multi-layer armor to shatter and absorb micrometeoroids.

322. **Q:** What is a terminator orbit (dawn-dusk)?

A: Satellite flies along the **day-night line** for constant lighting.

323. **Q:** What is a gravity well?

A: A visual metaphor for **gravitational potential** around a mass.

324. **Q:** What is Lambert albedo vs. geometric albedo?

A: Different **measures of reflectivity** under assumed scattering laws.

325. **Q:** What is a capture cross-section?

A: Effective area for a process like **collision or absorption**.

326. **Q:** What is a bolometer?

A: A detector that measures **radiant energy** via heating.

327. **Q:** What is extinction in astronomy?

A: Dimming by dust and gas between source and observer.

328. **Q:** What is scintillation?

A: Twinkling of starlight from atmospheric turbulence.

329. **Q:** What is a seeing disk?

A: Apparent **blurring** size from atmospheric seeing.

330. **Q:** What is the terminator illusion on the Moon?

A: Features look **sharper near sunrise/sunset** due to long shadows.

331. **Q:** What is libration of the Moon?

A: A **rocking motion** letting us see slightly more than half the surface over time.

332. **Q:** What is a selenographic coordinate?

A: Longitude/latitude system on the Moon.

333. **Q:** What is planetary occultation of a star good for?

A: Measuring **atmospheres** and **precise sizes**.

334. **Q:** What is the Tsiolkovsky rocket equation about?

A: How **delta-v depends on mass ratio** and exhaust velocity.

335. **Q:** What is staging in rockets?

A: Dropping **empty sections** to improve performance.

336. **Q:** What is throttling?

A: Adjusting **engine thrust** for control and efficiency.

337. **Q:** What is gimbaling?

A: Pivoting engines to steer thrust direction.

338. **Q:** What is roll, pitch, and yaw?

A: The three **rotation axes** of a spacecraft.

339. **Q:** What is a barycentric dynamical time (TDB)?

A: A uniform **time standard** used in precise ephemerides.

340. **Q:** What is an ephemeris?

A: A **table or model** of an object's position over time.

341. **Q:** What is light-time correction?
A: Adjusting for **finite light travel** when timing observations.

342. **Q:** What is a phase angle in photometry?
A: Angle Sun-object-observer; affects **brightness and glints**.

343. **Q:** What is specular reflection in space images?
A: Mirror-like glints from smooth surfaces (e.g., oceans, icy plains).

344. **Q:** What is a terminator shadow length used for?
A: Estimating **feature heights** on airless bodies.

345. **Q:** What is a tessera on Venus?
A: Highly deformed terrain suggesting complex tectonics.

346. **Q:** What is Maxwell Montes noteworthy for?
A: One of **Venus's highest mountain ranges**.

347. **Q:** What is the Caloris Basin?
A: A **huge impact basin** on **Mercury**.

348. **Q:** What is the Tharsis region?
A: A **volcanic plateau** on **Mars** with giant shield volcanoes.

349. **Q:** What are yardangs?
A: Wind-carved ridges seen on Mars and Earth deserts.

350. **Q:** What is aeolian activity?
A: Wind-driven processes shaping landscapes.

351. **Q:** What are dunefields on Mars made of?
A: Sand-sized grains, often basaltic.

352. **Q:** What is a brain terrain on Mars?

A: Ice-related **textured landforms** resembling convolutions.

353. **Q:** What is a ring current?

A: A **circulating current** of charged particles around Earth affecting geomagnetic storms.

354. **Q:** What is magnetotail?

A: The **night-side extension** of a planet's magnetosphere.

355. **Q:** What is reconnection in space plasmas?

A: Magnetic field lines break and reconnect, releasing energy.

356. **Q:** What is an L-shell (McIlwain)?

A: A parameter describing **magnetic field lines** by distance.

357. **Q:** What is a stellar nursery nickname for Orion Nebula?

A: M42, a lively **star-forming region**.

358. **Q:** What is Barnard's Star known for?

A: Very **high proper motion**, it crosses the sky quickly over decades.

359. **Q:** What is proper motion?

A: A star's **apparent motion** across the sky over time.

360. **Q:** What is an ecliptic?

A: The **Sun's apparent path**; plane of Earth's orbit.

361. **Q:** What is the celestial equator?

A: Earth's **equator projected** onto the sky.

362. **Q:** What are right ascension and declination?

A: Sky coordinates analogous to longitude/latitude.

363. **Q:** What is atmospheric seeing best when?

A: Stable, calm air, often after midnight, at high sites.

364. **Q:** What is light pollution?

A: Excess artificial light that reduces night-sky visibility.

365. **Q:** What is dark adaptation?

A: Eyes becoming **more sensitive** in low light after minutes.

366. **Q:** What is averted vision?

A: Looking slightly **off to the side** to see faint objects better.

367. **Q:** What is a planetary conjunction?

A: Two planets **appear close** in the sky.

368. **Q:** What is syzygy?

A: Three bodies aligned (e.g., during eclipses).

369. **Q:** What is a superior planet (classical)?

A: A planet whose **orbit is beyond Earth's** (Mars outward).

370. **Q:** What is an inferior planet (classical)?

A: A planet **inside Earth's orbit** (Mercury, Venus).

371. **Q:** What is an elongation (inner planets)?

A: Angular distance from the Sun; **greatest elongation** = best viewing.

372. **Q:** What is zodiac?

A: The **belt of sky** along the ecliptic where planets move.

373. **Q:** What is precovery in astronomy?

A: Finding **earlier images** of a newly discovered object.

374. **Q:** What is a light bucket?

A: Nickname for a **large-aperture** telescope.

375. **Q:** What is field of view (FOV)?

A: The **angular size** of sky an instrument can see at once.

376. **Q:** What is limiting magnitude?

A: The **faintest object** detectable under given conditions.

377. **Q:** What is a seeing-limited resolution?

A: Resolution **capped by atmosphere**, not telescope size.

378. **Q:** What is diffraction limit?

A: The **best possible** resolution for a given aperture and wavelength.

379. **Q:** What is interferometry?

A: Combining **signals from multiple telescopes** to act like a larger one.

380. **Q:** What is the Very Large Array (VLA) famous for?

A: Radio interferometry with movable dishes in New Mexico.

381. **Q:** What is a photometer?

A: An instrument measuring **light intensity** precisely.

382. **Q:** What is spectroscopy used for?

A: Splitting light to **identify elements**, temperatures, motions.

383. **Q:** What is Doppler broadening?

A: Line widening from motion/temperature of emitting gas.

384. **Q:** What is a standard candle?

A: An object of **known luminosity** used to measure distance.

385. **Q:** What is K-correction?

A: Adjusting **galaxy magnitudes** for redshifted spectra.

386. **Q:** What is a red dwarf's key trait?

A: Very long lifetimes, trillions of years.

387. **Q:** What is a star's metallicity notation [Fe/H]?

A: Logarithmic iron abundance relative to the Sun.

388. **Q:** What is an H II region?

A: Ionized hydrogen cloud around hot young stars.

389. **Q:** What is a P Cygni profile?

A: Spectral **emission + blueshifted absorption** indicating outflows.

390. **Q:** What is escape of atmospheres called?

A: Jeans escape (thermal) or **sputtering** (non-thermal), among others.

391. **Q:** What is planetary scale height?

A: The altitude where **pressure drops by ~e**.

392. **Q:** What is Rayleigh scattering?

A: Short-wavelength light scatters more, why **Earth's sky is blue**.

393. **Q:** What is Mie scattering?

A: Particle-size scattering, important in clouds/dust.

394. **Q:** What is opposition effect on asteroids?

A: Sharp brightening at low phase angles.

395. **Q:** What is a coma on a comet?

A: The **fuzzy atmosphere** around its nucleus.

396. **Q:** What is outgassing?

A: Volatiles escaping from a heated body (comet, planet).

397. **Q:** What is a planetary hot spot (atmospheric)?

A: A **warm region** due to dynamics or waves (e.g., Jupiter).

398. **Q:** What is baroclinic instability?

A: Atmospheric instability that forms cyclones/eddies.

399. **Q:** What is a superrotation (Venus)?

A: Atmosphere **circulates faster** than the planet rotates.

400. **Q:** What keeps us going back to space?

A: Curiosity, measurement, and the habit of doing hard things well.

The International Space Station

A practical, hopeful story about people learning to live together above the world

Look up on a clear night and you might see a bright point sliding across the sky. No blinking lights. No engine noise. Just a white bead on a black thread, crossing from west to east. That's the International Space Station, the ISS, a football-field-sized laboratory and home that circles Earth about every 90 minutes. It is the most complicated machine humans have ever built off the planet. It is also a promise kept: that countries can work together on something hard, risky, and long-term, and keep it running day after day.

The ISS is not a castle in the clouds. It's plumbing and pumps, laptops and lab racks, wires and wrenches, Velcro and handrails, filters and fans. It's also experiments, sunrises, and quiet moments where a crew member presses a forehead to a window and watches the blue curve of the Earth slide by. It is both ordinary and amazing. That's its magic.

1) Why Build a Station at All?

A space station answers a simple question: what do we learn by *staying*?

Short missions teach survival. Long missions teach living. If humanity wants to travel farther, back to the Moon for months, on to Mars for years, we must understand what time, weightlessness, and isolation do to bodies, minds, machines, and teams. We need a place to test life-support systems, practice repairs, try new tools, grow plants, and run thousands of experiments you can't do on Earth because gravity always gets in the way. The ISS is that place.

A second reason is cooperation. During the Cold War, the United States and the Soviet Union competed in space. Afterward, leaders asked a wise question: What if we turn competition into partnership? A shared station would keep engineers busy building, keep astronauts and cosmonauts working side by side, and keep doors open even when politics on Earth got noisy. The ISS became both a lab and a handshake.

2) Who Built It?

The ISS is an international project. The main partners are:

- **United States (NASA)**
- **Russia (Roscosmos)**
- **Europe (ESA-European Space Agency)**
- **Japan (JAXA)**
- **Canada (CSA)**

Each partner brings hardware, rockets, experiments, and people. The U.S. built key modules, the large solar arrays, and truss structure. Russia provided the first living quarters and propulsion to keep the station in orbit. Europe and Japan built big research labs. Canada built the station's famous robot arm. Many other nations contribute experiments. It's a group project where everyone actually did their part.

3) How Big Is It? Where Does It Orbit?

The ISS is about the size of a football field from end to end, including its solar panels. Inside, it has about as much pressurized volume as a large airplane, several connected modules forming rooms, hallways, and labs. It orbits at roughly 400 kilometers (250 miles) above Earth in low Earth orbit (LEO). At that height, it still feels a thin touch of the upper atmosphere, so it slowly loses altitude and must be "reboosted" regularly by spacecraft engines to keep flying.

The station circles Earth at about 7.66 kilometers per second (17,500 mph). That means **about 16 sunrises and 16 sunsets every day**. Time onboard is kept in Coordinated Universal Time (UTC) so crews and control centers around the world can sync up.

4) A Quick Timeline

- **1998**: First pieces launched, starting with Russia's **Zarya** module and the U.S. **Unity** node. Assembly begins.
- **2000**: First long-duration crew (Expedition 1) moves in. From then on, the ISS has been continuously inhabited.
- **2000-2011**: The Space Shuttle builds out the station, delivering truss segments, big solar arrays, labs, and supplies.
- **2011-present**: After the Shuttle retires, cargo and crew switch to other vehicles: Russia's Soyuz and Progress; Europe's ATV; Japan's HTV (now HTV-X); U.S. commercial craft like **SpaceX Dragon** and **Northrop Grumman Cygnus**; later, crewed **SpaceX Crew Dragon** and **Boeing Starliner**.

Assembly took more than a decade and over 100 spacewalks. The result is a strong, flexible platform for science and training.

5) What Are the Main Parts?

Think of the ISS as a long spine (the **truss**) with big **solar arrays** and **radiators** sticking out like wings, and a string of **pressurized modules** attached like beads on a necklace.

Pressurized Modules (Where People Work and Live)

- **Zarya** (Russia): First module; provides storage and initial power/propulsion interfaces.

- **Zvezda** (Russia): Early living quarters; includes sleeping area, galley, and life support.
- **Unity / Node 1** (U.S.): The first U.S. connection node.
- **Destiny** (U.S.): Main U.S. laboratory.
- **Harmony / Node 2** (U.S.): Connector with docking ports for visiting vehicles.
- **Tranquility / Node 3** (U.S.): Houses life support, exercise gear; attached **Cupola** window module for robotics and Earth views.
- **Columbus** (Europe): European lab with experiment racks.
- **Kibo** (Japan): Japanese lab, includes a pressurized module, **logistics module**, and an **external platform** for experiments in open space.
- **Rassvet, Poisk, Nauka** (Russia): Russian research and docking modules; **Nauka** adds lab space and a new European robot arm (ERA).
- **BEAM** (U.S., Bigelow): An **inflatable** test module, small, light, expanded in orbit to test future habitat ideas.

The Truss and Power

- **Integrated Truss Structure**: Long backbone holding power equipment and thermal control.
- **Solar Arrays**: Giant panels that convert sunlight to electricity.
- **Radiators**: Panels that dump heat to space because in vacuum, radiation is the only way to cool.

Robotics

- **Canadarm2 (Canadarm-two)**: A long, multi-jointed robotic arm that "walks" around the station by grabbing fixtures. Used to capture cargo ships, move equipment, and assist spacewalkers.

- **Dextre**: A two-armed "robotic handyman" that can change instruments and do fine tasks to avoid some risky spacewalks.

- **ERA (European Robotic Arm)**: Works on the Russian segment, supports tasks around Nauka and future Russian modules.

Attitude Control

- **CMGs (Control Moment Gyroscopes)** spin like tops and, when gimbaled, twist the station without using fuel. For bigger pushes or reboosts, visiting spacecraft provide thruster power.

6) How Do Vehicles Dock or Berth?

Visiting vehicles use two main methods:

- **Docking**: A spacecraft flies itself (or is guided) to attach directly to a docking port. Soyuz, Progress, Crew Dragon, and Starliner dock.

- **Berthing**: A spacecraft parks close, then the arm (Canadarm2) reaches out, captures it, and moves it to a port. Cygnus and older cargo Dragons used berthing; newer Dragons now **dock**.

Ports are standardized so different craft can use them safely. Crews perform leak checks before opening hatches. Cargo gets unloaded, trash gets loaded, and then the ship departs, often burning up in the atmosphere to dispose of waste, or returning with science samples (as Dragon does).

7) How Do People Live Up There?

Water is precious. The station recycles as much as possible. Moisture from breath and sweat, and even purified urine, goes back into clean water. A joke says, "Yesterday's coffee becomes tomorrow's coffee." It's funny because it's true and it works.

Air is managed by machines that scrub carbon dioxide and add oxygen. **Temperature** and **humidity** are controlled by fans and heat exchangers. There is no "up" or "down," so everything has Velcro or bungee straps. Food arrives shelf-stable in packets. Crew members heat it, add water if needed, and eat with a spoon or fork that won't float away.

Sleep happens in small crew cabins the size of a phone booth with a sleeping bag tethered to the wall. Lights are dimmed according to a daily schedule, not the sun, because the sun rises every 90 minutes and no one wants jet lag sixteen times a day.

Exercise is not optional. In weightlessness, muscles and bones weaken. Astronauts work out about **two hours a day** using a treadmill, a resistance device that mimics weightlifting, and a stationary bike. The goal is to return to Earth with bones, hearts, and balance still strong.

Hygiene uses rinseless wipes and no-rinse shampoo. Haircuts happen with a vacuum clippers. The bathroom uses airflow to do what gravity does on Earth. It's not glamorous; it's engineering.

8) What Kind of Science Happens on the ISS?

The ISS is a factory for experiments that need **microgravity**, a near-weightless environment, plus steady power, data, and a crew who can tweak and fix things.

Human Health and Biology

- **Bone and Muscle Loss**: Measure how fast it happens and how to slow it.
- **Fluids and Vision**: Fluids shift upward in microgravity, sometimes affecting eyesight. Scientists study why and how to protect crews.
- **Immune System**: Changes in space can alter how the body fights germs. Understanding this helps both astronauts and patients on Earth.
- **Plants in Space**: Lettuce, radishes, dwarf wheat, learning how to grow food off-planet.

Physics and Materials

- **Combustion**: Flames behave like gentle blue spheres in microgravity. Studying them can improve fire safety and engines.
- **Fluids**: Without gravity, liquids form floating globes and behave in surprising ways. This helps design fuel tanks and water systems for future craft.
- **Colloids and Alloys**: Making materials without sediment or convection can reveal new structures and properties.

Technology Demos

- **Life Support Systems**: Testing better CO_2 scrubbers, air filters, and water recyclers.
- **3D Printing**: Printing tools and parts in space to reduce resupply.
- **Cold Atom Experiments**: Chilling atoms to near absolute zero to study quantum physics without Earth's vibrations.

Earth and Space Observations

- **Earth Imaging**: Tracking storms, fires, glaciers, crops, and coastlines.
- **Cosmic Ray and Particle Detectors**: Learning about radiation in space and fundamental particles.

The station is like a city block of labs with a permanent crew. If something breaks, they don't call a building manager, they *are* the building manager.

9) Spacewalks: The Work Outside

A **spacewalk** (EVA, extravehicular activity) is when crew members don a suit, depressurize an airlock, and step outside. They mount hand over hand along the station's exterior using tethers and handrails. Tasks include installing new instruments, fixing pumps, routing cables, and swapping out batteries. Suits are miniature spaceships: they provide oxygen, remove CO_2, regulate temperature, and carry a small jet "safety pack" in case the astronaut drifts away (rare, and well-planned against).

Spacewalks are choreographed like dances. Crews practice underwater on Earth in giant pools called neutral-buoyancy labs. On the day, mission control reads steps, the crew follows, and the station slowly rolls across a patchwork of continents and seas while work gets done one bolt at a time.

10) How Does the Station Keep Its Balance and Stay Put?

The ISS must keep its **attitude** (orientation) for power, communications, and safety. The **CMGs**, spinning gyroscopes, tilt and twist the station without burning fuel. When gyros saturate (reach their limits) or when bigger moves are needed, **thrusters** on visiting spacecraft nudge the station.

To keep its **altitude**, the station performs **reboosts** every so often. Engines on attached ships push it to a higher orbit. The station also performs **debris avoidance maneuvers** when tracking shows something might come too close. Most space junk passes harmlessly. Sometimes the crew even shelters in their docked spacecraft for a short time if a piece zips near and a maneuver can't be done in time. Procedures are practiced. Safety is culture.

11) Communications: How Do We Talk to the ISS?

The station talks to Earth mostly through **relay satellites**, for the U.S., the TDRS network, sending data, video, voice, and telemetry to mission control. Russia uses its own links. Europe, Japan, and Canada tie into the system through agreements. There's also **amateur radio** on the ISS, where crew members sometimes chat with students. Imagine answering a question in a classroom, then looking down and seeing the school a few minutes later as the station passes overhead. That's outreach with a pulse.

12) Who Runs the Day?

A typical day is packed but predictable:

- **Wake**: Crew checks messages, grabs breakfast, does a morning planning conference with control centers.
- **Work Blocks**: Science tasks, maintenance, cargo transfers, exercise, training, outreach events, more science.
- **Meals**: Quick, but social. Food is culture, even in microgravity.
- **Evening**: Finish reports, personal time, look out the window (no shame), prep for tomorrow.
- **Sleep**: Lights dim to match a stable schedule.

Flight controllers on Earth work in shifts. Specialists monitor life support, power, navigation, robotics, communications, and science. The crew follows a digital schedule that updates down to the minute. Flexibility matters; hardware doesn't always care about schedules.

13) What Did the ISS Cost? What Did It Buy?

The ISS cost many tens of billions of dollars across all partners. People sometimes ask if it was "worth it." A fair answer looks beyond one number.

What it bought:

- A **continuous human presence** in space for decades.
- A **tested playbook** for long missions: how to eat, sleep, work, exercise, and stay healthy off Earth.
- A **global supply chain** of high-reliability parts and people who know how to make them.
- **Science** that couldn't be done on Earth and technology demos that will make future missions safer and cheaper.

- A **cooperation template** that kept countries talking and working even when politics on the ground got cold.

You may not see all the returns on a ledger today. You see them in the missions that follow, the companies that learned to build for space, and the students who saw the station and decided to join the story.

14) Mishaps, Lessons, and the Value of Boring Excellence

Space is not a theme park. The ISS has faced air leaks, faulty pumps, stuck valves, ammonia issues, computer hiccups, and the constant threat of micrometeoroids. Each problem becomes a case study: detect early, isolate, work the checklist, fix safely. There have been dramatic days, but most days are steady and that steadiness is victory. "Boring excellence" is what keeps people alive and programs on track.

Two deep lessons:

1. **Listen to bad news.** Reward the person who raises a concern; they help you find the truth.
2. **Test like you fly.** Simulate the hard parts. Practice failures. Practice speaking up.

These rules were paid for in the past with tragedy. They are honored on the station every day.

15) Culture in Orbit: Ordinary Life in an Extraordinary Place

Culture is the unofficial oxygen of the station. Crews share meals across languages, swap music, bring national foods for special days, trade jokes, and teach each other phrases. Small rituals matter: a photo over a favorite city, a note left on a laptop, a holiday call from orbit to a classroom.

There's science and schedule, yes but also humanity. The station only works when people do.

The **Cupola**, a windowed dome, might be the most beloved spot. From there, crew members take photos and also use Canadarm2 with a clear view. Watching Earth from the Cupola changes people. You see the thinness of the atmosphere, the light of cities, the line of a river from mountain to sea. Borders are hard to spot from orbit. Storms are not. Many astronauts say the view makes them protective of the home they're floating above.

16) Education and Inspiration

The ISS is a long-running live show for science education. Students design simple experiments, watch live link-ups, track the station overhead, and sometimes talk by ham radio to the crew. Teachers use ISS photos in geography and earth science. Engineers use station problems as classroom cases: "Here's the issue, how would you fix it?" Inspiration cannot be graphed, but it shows up when a student picks a major, joins a lab, or starts a company.

17) The ISS and the Space Economy

The station helped grow a market for:

- **Cargo delivery** (multiple providers competing, good for price and schedules).
- **Crew transportation** (commercial crew changed the game).
- **In-space research services** (companies run racks and provide experiment support).
- **Earth observation** testbeds.

- **In-space manufacturing** demos (fibers, crystals, bioprinting prototypes).

Government became a **customer**, not the only builder. That shift invites efficiency and innovation. Not every idea will pay off, but the road is open now.

18) Health and Radiation: The Hard Problems

Two big risks in space are **radiation** and **long-term weightlessness**.

- **Radiation**: The ISS is protected by Earth's magnetic field, but crews still receive higher doses than on Earth. Shields and smart planning reduce exposure. Medical teams track long-term health carefully.

- **Weightlessness**: It weakens bones and muscles, can change vision, and alters fluid balance. Exercise, nutrition, and medical monitoring help. The point isn't to make space easy; it's to make it manageable.

These lessons are tickets to Mars. You don't buy them once. You keep paying attention.

19) Ethics and the Orbit We Share

Space is a commons. The ISS made the idea concrete. Partners share responsibilities: don't create debris, share traffic information, publish docking standards, and help each other when something goes wrong. These ideas show up in agreements and day-to-day habits. They may sound boring. Good. Boring rules prevent exciting disasters.

20) What Comes Next?

The ISS won't last forever. Metal tires, seals age, and orbits are not eternal. When its service ends, the partners plan a **controlled deorbit** over remote ocean and a careful handoff to **next-generation stations**, some government-built, some commercial, likely smaller and more specialized. Lessons from the ISS will flow into them:

- Better **life support** and recycling.
- **Modular** designs that swap in new tech.
- Smarter **robotics** and more **automation**.
- More **commercial services** and international users.
- Stronger focus on **science that pays forward** to deep-space missions and Earth benefits.

The ISS taught us how to live in orbit for decades. The next stations will teach us how to **thrive** there, cheaper, safer, and with more people involved.

21) A Tour Through a Workday Experiment (An Example)

Imagine you're the principal investigator (PI) for a plant growth study.

- **Design**: On Earth, you build a growth chamber that controls light, humidity, and nutrients. You make it compact, safe, and easy to operate with gloved hands.
- **Ride**: It launches inside a cargo ship, damped and strapped.
- **Install**: The crew unpacks the box, wires it to power and data, checks airflow, and loads seeds in a cartridge.

- **Run**: Cameras watch roots and leaves, sensors log CO_2 and moisture. A command from your console on Earth adjusts light levels.
- **Sample**: Crew collects leaves for later genetic study, places them in a tiny freezer, and logs times precisely.
- **Return**: A Dragon capsule splashes down with your samples. Back in your lab, you compare gene expression to Earth controls. You learn which switches in a plant flip when gravity vanishes. That helps both space farming and drought-resistant crops at home.

Multiply that by hundreds of experiments across biology, physics, medicine, and engineering. That's the station's heartbeat.

22) The Window Seat: Earth Science from 400 Kilometers Up

From orbit, you can watch **wildfires, hurricanes, algae blooms, dust storms, volcanic plumes**, and **ice loss**. Some instruments measure greenhouse gases; others scan oceans or forests. The ISS passes over the same regions again and again, a perfect platform for **time-series** data. Astronauts also take handheld photos guided by scientists, sometimes a person with a good eye beats an automatic camera.

This is not politics; it's measurement. When rivers rise and glaciers shrink, when reefs bleach and forests recover, the numbers help people prepare and plan. The view is beautiful. The data is useful.

23) Robots, Arms, and the Art of Not Going Outside

Every minute a person is outside, risk rises. That's why the station's robot arms matter. **Canadarm2** can "walk off" between fixtures and move like a slow, steady crane. **Dextre** turns bolts, removes boxes, and swaps instruments too delicate for space gloves. On the Russian side, the **ERA** supports tasks around Nauka. The dream is to do more maintenance by robot and fewer long spacewalks. Robots don't get tired. They also don't smile at the sunrise. So we keep both.

24) Power, Heat, and How Not to Freeze or Fry

The station makes electricity from sunlight and dumps heat through radiators. Managing heat is half the job in space. In sunlight you can bake; in shadow you can freeze. On Earth, air and water carry heat away. In space, only **radiation** works. That's why those big white panels matter. Inside, fans keep air moving so hot spots and cold corners don't form. Sensors watch everything. When a pump stumbles, the crew and ground team notice fast.

25) The Sound of the Station

The ISS is not silent. Fans hum. Pumps murmur. Laptops click. Velcro rips. The crew talks calmly. It sounds like a well-run workshop. That constant noise can tire you out, so crew members use earplugs at times. Sleep stations insulate against the hum. Silence is rare. But when you press your helmet to a window during a night pass, and the Earth glow is the only light, your brain finds a quiet of its own.

26) Safety, Redundancy, and the Rule of Three

Space hardware is built with **redundancy**, two or three of everything important. If one fails, another takes over. Lines are cross-connected. Valves can be rerouted. Fire sensors watch. Leak detectors listen. The crew trains for **fire**, **depressurization**, and **toxic leaks**. They drill. They label. They clean. Good housekeeping is safety, too: a floating bolt can jam a fan; a loose bag can block a vent. Order is kindness in orbit.

27) The Human Story: Firsts, Records, and Quiet Milestones

There are famous moments: first long-duration crews, first spacewalks from new modules, the first all-woman spacewalk, the first private crewed vehicles docking. There are also quiet milestones: a crew member running a marathon on the treadmill while a real race happens on Earth; a child asking a question by radio and hearing an answer from above; a plant bending toward LED light in a clear box.

Records, time in space, number of spacewalks, number of orbits, are nice. The **continuity** is nicer: that for over two decades, you could look up and know someone was up there, keeping watch and doing work.

28) What the ISS Taught Us (Short List)

- **Microgravity changes everything**, so test everything.
- **Exercise saves bodies**; culture saves teams.
- **Robotics reduce risk**; practice reduces panic.
- **Closed-loop life support works** (and can work better).
- **International teamwork can survive arguments** if the job matters and the rules are clear.
- **Commercial partnerships** can deliver steady rides and lower costs.
- **Earth is fragile and connected**; you feel that in your bones after six months at the window.

29) The Road Beyond the Station

The ISS is not the end. It's the bridge. Next steps include:

- **Lunar orbit stations** (a small "Gateway") to support Moon missions.
- **Surface bases** on the Moon to test long-term stays and resource use (like mining ice for water and fuel).
- **Mars-class life support** tested first in low Earth orbit and cislunar space.
- **Commercial stations** that rent lab racks or host short-term visits.
- **More countries and more students** taking part through cheaper access and better online tools.

The orbit is a workshop. The solar system is the job site.

30) Closing: A Lighthouse Above the Weather

The International Space Station is a lighthouse that circles the world. It doesn't blink warnings; it whispers possibilities. It says: we can build together, live together, solve problems together, and keep promises longer than a news cycle. It says: the view from up here will make you love down there more. It says: tradition, discipline, craft, and courage, can carry us into tomorrow without losing our balance.

Next time it passes over, step outside. Watch that bright dot cut a steady path. Somewhere on board, a plant is growing, a pump is humming, a laptop is logging data, a wrench is floating, a crew member is pedaling, and a camera is catching yet another sunrise. Sixteen of them today. Enough for everyone.

100 Influential Pioneers in Space Exploration

1. **Konstantin Tsiolkovsky** — Rocket equation pioneer; imagined multi-stage rockets, stations, and living in orbit.
2. **Robert H. Goddard** — Launched the first liquid-fuel rocket; invented guidance, pumps, and gyros for modern rocketry.
3. **Hermann Oberth** — Theorist who inspired Europe's early rocket societies; bridged dreamers and engineers.
4. **Wernher von Braun** — Led Saturn V development; controversial origins, transformative Apollo results.
5. **Sergei Korolev** — Soviet "Chief Designer"; drove Sputnik, Gagarin's flight, and R-7 launchers.
6. **Yuri Gagarin** — First human in space (1961); one orbit that changed history.
7. **Valentina Tereshkova** — First woman in space (1963); proved space isn't boys-only.
8. **Neil Armstrong** — First human on the Moon; test pilot calm, history's quiet line.
9. **Buzz Aldrin** — Apollo 11 moonwalker; champion of orbital rendezvous and Mars advocacy.
10. **Michael Collins** — Apollo 11 CMP; kept the ride home perfect while the world watched the surface.

11. **John Glenn** — First American to orbit; later returned to space as elder statesman.
12. **Alan Shepard** — First American in space; Apollo 14 moonwalker with lunar "golf."
13. **Virgil "Gus" Grissom** — Mercury/Gemini veteran; Apollo 1 loss reshaped NASA safety culture.
14. **Alexei Leonov** — First spacewalk; Apollo–Soyuz handshake made cooperation real.
15. **Gene Kranz** — Flight director; "tough and competent" creed and Apollo 13 leadership.
16. **Chris Kraft** — Architect of Mission Control; procedures, roles, and calm under pressure.
17. **Katherine Johnson** — Mathematician who computed Mercury/Apollo trajectories; opened doors for many.
18. **Margaret Hamilton** — Led Apollo flight software; coined "software engineering"; code saved landings.
19. **Mary W. Jackson** — NASA aeronautical engineer; turned wind-tunnel data into safer flight.
20. **Dorothy Vaughan** — Math/computing lead; guided transition from human calculators to machines.
21. **George Mueller** — Managed Apollo "all-up" testing; integrated the moonshot supply chain.
22. **Maxime Faget** — Designer behind Mercury, Gemini, early Shuttle shapes.

23. **Robert Gilruth** — Led Space Task Group and Manned Spacecraft Center; built the teams.
24. **John Houbolt** — Championed lunar-orbit rendezvous; changed the Moon plan.
25. **Rocco Petrone** — Apollo launch chief; master of countdown discipline.
26. **Jim Lovell** — Apollo 13 commander; calm decisions turned failure into survival.
27. **Frank Borman** — Apollo 8 commander; first lunar orbit, "Earthrise" perspective.
28. **Gene Cernan** — Last Apollo moonwalker (so far); urged a mindful return.
29. **Pete Conrad** — Apollo 12 commander; precision lunar landing by design.
30. **John Young** — Flew Gemini, Apollo, first Shuttle; spanned eras of crewed flight.
31. **Sally Ride** — First American woman in space; lifelong STEM champion.
32. **Mae Jemison** — First Black woman in space; physician-engineer-educator.
33. **Peggy Whitson** — U.S. record-holder for time in space; ISS commander and science leader.
34. **Eileen Collins** — First woman Shuttle pilot and commander; precise leadership.

35. **Svetlana Savitskaya** — Second woman in space; first woman to perform a spacewalk.

36. **Charles Bolden** — Astronaut and NASA Administrator; advanced commercial crew and science.

37. **Edward C. Stone** — Voyager chief scientist; guided decades of interstellar exploration.

38. **Carolyn Porco** — Cassini imaging lead; turned Saturn's system into a living story.

39. **Carl Sagan** — Planetary scientist, explainer, Voyager Golden Record; made the cosmos feel close.

40. **Jack Parsons** — JPL co-founder; early solid-fuel propulsion pioneer.

41. **William H. Pickering** — JPL director; Explorer 1, Ranger, early planetary missions.

42. **Qian Xuesen (Tsien Hsue-shen)** — Architect of China's missile/space program; systems thinker.

43. **Hideo Itokawa** — Father of Japanese rocketry; groundwork for sample-return missions.

44. **Vikram Sarabhai** — Father of India's space program; "space for society" vision.

45. **A.P.J. Abdul Kalam** — Indian launch-vehicle leader and president; built reliable access to space.

46. **Sergei Krikalev** — Endurance cosmonaut; rode out USSR's collapse in orbit; helped build ISS.

47. **Valeri Polyakov** — Longest single spaceflight (14+ months); data on living long in space.

48. **Elon Musk** — SpaceX co-founder; made reusability routine and cut launch costs.

49. **Gwynne Shotwell** — SpaceX president/COO; turned bold plans into dependable cadence.

50. **Jeff Bezos** — Blue Origin founder; funded long-horizon reusability and suborbital human flights.

51. **Edwin Hubble** — Proved galaxies beyond the Milky Way; discovered cosmic expansion.

52. **Henrietta Swan Leavitt** — Found Cepheid period–luminosity law; cornerstone distance yardstick.

53. **Annie Jump Cannon** — Built OBAFGKM classification; organized stellar zoo for physics.

54. **Cecilia Payne-Gaposchkin** — Showed stars are mostly hydrogen/helium; reset stellar chemistry.

55. **Vera Rubin** — Galaxy rotation curves; revealed dark matter's gravitational hand.

56. **Jocelyn Bell Burnell** — Co-discovered pulsars; unveiled neutron star beacons.

57. **Subrahmanyan Chandrasekhar** — White-dwarf mass limit; mapped stellar fates with math.

58. **Arthur C. Clarke** — Geostationary satellite concept; visionary who became infrastructure.

59. **James Van Allen** — Discovered radiation belts; defined near-Earth space hazards.

60. **Harold Rosen** — Built first practical GEO comsat (Syncom); global comms backbone.

61. **John R. Pierce** — Led Telstar-era satellite communications at Bell Labs.

62. **Gerard K. O'Neill** — Proposed rotating space habitats; "live in space" reframing.

63. **Freeman Dyson** — Big, testable ideas (propulsion, Dyson spheres); expanded what's thinkable.

64. **Burt Rutan** — Designed SpaceShipOne; proved small teams can fly crewed space.

65. **Peter Diamandis** — Launched XPRIZE; ignited private human-spaceflight competition.

66. **Richard Branson** — Backed Virgin Galactic; carried space tourism into reality.

67. **Peter Beck** — Rocket Lab founder; normalized reliable smallsat launch.

68. **Hans Koenigsmann** — Early SpaceX reliability chief; from prototypes to dependable ops.

69. **Tom Mueller** — Merlin/Raptor engine architect; modern reusable thrust.

70. **Tory Bruno** — ULA CEO; pushed Vulcan and steady national-security launch.

71. **Nancy Grace Roman** — "Mother of Hubble"; championed space observatories inside NASA.

72. **James E. Webb** — NASA Administrator; built the management engine behind Apollo.

73. **John C. Mather** — COBE Nobel; JWST senior scientist; infrared universe opener.

74. **George F. Smoot** — COBE co-lead; mapped primordial CMB ripples.

75. **Rainer Weiss** — LIGO co-founder; turned gravitational waves into data.

76. **Kip Thorne** — Gravity theorist/LIGO architect; bridged deep physics and detection.

77. **Andrea Ghez** — Proved Milky Way's central black hole via stellar orbits.

78. **Reinhard Genzel** — Parallel Sgr A* proof; independent confirmation of the giant within.

79. **Eugene Shoemaker** — Founded planetary geology; linked impacts on Earth to cosmic events.

80. **Carolyn Shoemaker** — Prolific comet-hunter; co-found of Shoemaker–Levy 9 impact on Jupiter.

81. **Mike Brown** — Discovered Eris and big KBOs; forced Pluto's reclassification.

82. **Alan Stern** — New Horizons PI; revealed Pluto's complex, active world.

83. **Charles Elachi** — JPL director/radar pioneer; steered Cassini, Mars fleet, Earth missions.

84. **Thomas Zurbuchen** — NASA science chief; greenlit daring missions, shepherded JWST.

85. **Chris Hadfield** — Canadian ISS commander; mastered outreach and operations.

86. **Samantha Cristoforetti** — ESA astronaut/ISS commander; science ops + public engagement.

87. **Tim Peake** — British ESA astronaut; ISS mission as national STEM rally.

88. **Scott Kelly** — Year-in-space twin study; cornerstone data for long-duration health.

89. **Mikhail Kornienko** — Kelly's year-long crewmate; endurance partner in science.

90. **Oleg Kononenko** — Record cumulative time in space; symbol of steady professionalism.

91. **Christina Koch** — Longest single spaceflight by a woman; Artemis II crewmember.

92. **Jessica Meir** — Biologist-astronaut; first all-woman spacewalk with Koch.

93. **Sunita Williams** — Veteran astronaut/ISS commander; spacewalk ace, commercial crew flier.

94. **Kalpana Chawla** — Columbia crewmember; legacy deepened safety culture, inspired millions.

95. **Yang Liwei** — China's first taikonaut; Shenzhou 5 proved independent human spaceflight.

96. **Liu Yang** — First Chinese woman in space; milestone for China's crewed program.

97. **Wang Yaping** — Taikonaut-teacher; live lessons from orbit to millions.

98. **Rakesh Sharma** — First Indian in space; made space part of India's national story.

99. **Alexander Gerst** — German ESA astronaut/ISS commander; geophysicist-eye Earth advocate.

100. **Marc Garneau** — First Canadian in space; later national leader keeping space in policy.

How AI Has Affected Space Exploration

Space used to be all checklists and radio calls. It still is, but now the checklists think with us. Artificial intelligence, by which we mean a toolbox of machine learning, pattern recognition, optimization, and increasingly capable language and vision systems, has slipped into nearly every corner of the space game. AI plans launches, drives rovers, keeps spacecraft healthy, sorts science from noise, helps astronauts juggle tasks, steers antennas, scouts hazards, and spots the needle in a haystack of stars.

In a sentence: **AI turns raw data into decisions faster than people can, and it keeps working while we sleep, while we're out of signal, and while the universe throws curveballs.**

1) Launch and Manufacturing - Smarter Engines, Fewer Scrapped Parts

Before a rocket ever meets a launch pad, AI is already at work.

- **Design optimization:** Engineers use learning-based optimizers to search huge design spaces, nozzle shapes, tank structures, injector patterns, finding options that balance performance, weight, cost, and safety. What used to take months of manual iteration can be explored in days.
- **Predictive manufacturing:** Machine learning watches **additive manufacturing** (3D printing of metal parts) in real time. It reads sensor traces, video, and thermal maps to catch defects before they grow into cracks. That means fewer scrapped engines, lower costs, and more consistent quality.

- **Test analytics:** Hot-fire tests create rivers of data, pressures, temperatures, vibrations. AI models learn the "normal song" of a healthy engine and flag off-key notes that humans might miss. Small anomalies spotted early can prevent big failures later.

- **Launch weather and windows:** Forecasting for launch is tricky. AI blends satellite images, radar, and historical patterns to recommend go/no-go windows with clearer risk estimates. The sky is still the boss, but we read it with sharper eyes.

Bottom line: AI trims fat, spots faults, and turns "good enough" into "we can do better."

2) Mission Design and Scheduling - Where Algebra Meets Orchestra

Space missions are moving puzzles: propellant limits, planetary alignments, communication passes, crew day/night cycles, conflicting experiments, and the occasional surprise. AI helps untangle the knots.

- **Trajectory search:** Hybrid optimizers explore thousands of interplanetary routes, gravity assists, and burn schedules. They trade fuel against time and risk, surfacing options a human planner might never try.

- **Deep Space Network (DSN) scheduling:** Dozens of missions want the same big antennas. AI schedulers negotiate constraints and produce fair, efficient timetables, then adjust when a spacecraft cries "urgent."

- **On-orbit timelining:** On the ISS and future stations, AI helps build daily plans that respect crew health limits, power/thermal budgets, and payload needs. If a pump hiccups at noon, the plan reflows automatically: swap tasks, slide experiments, keep the day productive.

Bottom line: AI plays conductor so the orchestra can actually play.

3) Autonomy Onboard - When "Out of Range" Doesn't Mean "Out of Luck"

Mars is minutes away by radio. The outer planets are hours. You can't joystick a rover through that kind of delay. Autonomy is the difference between **waiting** and **working**.

- **Guidance, Navigation, and Control (GNC):** Star trackers, cameras, and inertial sensors feed AI-driven estimators. Spacecraft recognize star fields, landmarks, and horizon lines; they refine their own navigation even when ground support is sleeping.
- **Terrain Relative Navigation (TRN):** Landers compare live camera views to onboard maps to avoid boulders and cliffs. That's computer vision plus fast decision-making, shaving risk at the scariest moment of a mission: descent.
- **Onboard science triage:** Rovers and orbiters capture more data than they can downlink. AI ranks images and spectra by "interestingness" (new rocks, unusual clouds, dust devils), pushing the best stuff first. Scientists back on Earth wake to highlights instead of haystacks.
- **Fault detection, isolation, and recovery (FDIR):** Anomalies start quietly. AI models learn the normal heartbeat of valves, gyros, batteries, and radios. When a reading drifts, the system can isolate the fault and sometimes reconfigure around it, switch to a

backup, reduce load, or change attitude, while the ground writes a longer-term fix.

Bottom line: Autonomy keeps robots useful when silence is normal and delays are long.

4) Surface Exploration - Rovers That Read the Road

Planetary surfaces are messy. Rocks, dust, slopes, shadows, every step is a question. AI helps rovers answer those questions without waiting for home.

- **Hazard detection:** Computer vision marks sand traps and sharp rocks; path planners choose routes that balance safety and speed. The rover still moves slowly, but it moves **daily**, even between downlinks.

- **Science targeting:** Rock textures, layers, and colors hint at past water or volcanic activity. AI helps choose targets for drills and spectrometers, saving human time for the subtle calls.

- **Aerial scouts:** Helicopters and drones on other worlds rely on visual odometry and learned terrain features to stabilize flight, pick landing zones, and explore places a rover can't reach. Their images feed rover planners with fresh context.

- **Sample caching strategy:** When a mission needs to collect and store rock cores for later pickup, AI tracks sample diversity, mineral types, formations, locations, to avoid ten copies of the same story.

Bottom line: The mobility bottleneck shifts from "we can't see" to "we can choose well."

5) Astronaut Assistants - A Second Brain with Good Manners

On stations and future habitats, time is precious and risks are non-zero. AI becomes a quiet crew member.

- **Procedure guidance:** Natural-language systems read a request like, "Show me the leak-check steps," and display the right checklist with diagrams, timers, and cautions. Hands stay gloved; minds stay clear.
- **Inventory and logistics:** Cameras plus object recognition track tools and supplies. "Where's wrench 3?" is no longer a 30-minute float-around question. The system suggests restock priorities for the next cargo ship.
- **Schedule support:** AI balances crew work, sleep, exercise, and comm windows. If a repair runs long, it reschedules two lower-priority tasks and warns a researcher on Earth about a delay.
- **Safety watch:** Wearables stream heart rate, sleep quality, and CO_2 exposure. Models flag fatigue risk or early signs of motion sickness, recommending adjustments before performance drops.
- **Robotic helpers:** Free-flying bots handle inventory scans, environmental sampling, and camera work. Think of them as floating tripods with a brain.

Bottom line: AI turns "where did I put that?" and "what's next?" into solved problems.

6) Keeping Spacecraft Healthy - Maintenance Before Emergency

Hardware in space can't visit the shop. Predicting trouble beats heroic repairs.

- **Telemetry guardians:** Thousands of channels stream temperatures, voltages, vibrations. AI spots subtle patterns, a pump starting to run warm at a certain attitude; a battery aging faster after many fast charge cycles.

- **Predictive maintenance:** If a valve tends to stick after a specific thermal cycle, the system recommends an earlier lube or a gentler mode. For stations and long cruises, these small saves add up to months of extra life.

- **Radiation and power budgeting:** Models adapt operations to solar storms and eclipse seasons. They throttle experiments, reshuffle charging plans, and protect sensitive instruments without human micromanagement.

Bottom line: "An ounce of prediction" beats "a pound of improvisation."

7) Science Firehose - From Raw Bits to New Worlds

The cosmos is loud with data. AI keeps up.

- **Exoplanets:** Transit surveys watch hundreds of thousands of stars. AI filters the tiny dips that mean "planet," rejects instrumental ghosts, and helps rank promising targets for follow-up. It also analyzes **radial velocity** time series to separate true signals from stellar noise.

- **Transient astronomy:** The sky changes, supernovae, gamma-ray bursts, gravitational-wave follow-ups. AI scans nightly images, flags new points of light, cross-matches catalogs, and alerts telescopes to swing fast. Minutes matter.

- **Galaxy mapping and morphology:** Deep nets classify galaxies by shape, measure redshifts from spectra, and trace the cosmic web across huge surveys. What once took years of hand-labeling now takes hours of training and minutes of inference.

- **Planetary science:** AI counts craters, maps dunes, segments ice cliffs, and finds fresh landslides on Mars or moons. It builds mineral maps from hyperspectral cubes, thousands of wavelengths per pixel, so geologists can ask better questions.

- **Astrobiology hints:** In chemical data, AI hunts for unusual patterns, molecules out of balance, spectral lines in odd ratios, that might be biosignatures to test more carefully. It doesn't declare life; it **prioritizes the hunt**.

Bottom line: AI is the tireless grad student who never sleeps and always labels perfectly.

8) Earth From Orbit - Better Eyes for a Shared Home

While "space exploration" often means the Moon and planets, much of it is looking back at Earth to understand and protect home. AI supercharges that too.

- **Cloud and aerosol detection:** Models clean satellite images and retrieve clearer signals for climate and weather research.

- **Disaster response:** AI spots fires, floods, storm damage, and oil spills from space and routes alerts to responders fast. Resolution improves; latency drops; help arrives sooner.

- **Cryosphere watch:** Algorithms measure glacier flow and sea-ice change, comparing weekly or daily images to decades of history.

- **Agriculture and water:** AI estimates soil moisture, crop health, and drought risk, guiding smarter irrigation and storage.

Bottom line: The best way to explore space responsibly is to keep our launchpad in good shape.

9) Space Traffic and Debris - Collision Avoidance in a Crowded Sky

Low Earth orbit is busy. So is geostationary orbit. Avoiding collisions takes constant attention.

- **Object detection:** Radar and optical sensors feed AI that tracks small, faint debris. Better detection means better catalogs.
- **Conjunction assessment:** Models predict close approaches days ahead, assess collision probabilities, and recommend maneuvers that burn little fuel and don't create new problems.
- **Autonomous satellite response:** Onboard systems can execute pre-cleared avoidance burns when ground control can't respond in time, then report back.
- **End-of-life planning:** AI helps design deorbit plans that minimize risk and comply with guidelines, no "space littering."

Bottom line: Good neighbors share data, and smart satellites dodge politely.

10) Compression, Comms, and Edge AI - Smarter Bits, Smaller Bills

Downlink is precious. Not everything can come home.

- **Onboard compression and summarization:** AI encoders reduce data size without losing the parts scientists care about. For images, that can mean keeping sharp edges and spectral features intact while tossing background redundancy.
- **Event-triggered recording:** Instruments don't have to run full-speed all the time; they can spike to high-rate modes when AI hears or sees something unusual.
- **Edge chips:** Low-power AI processors fly on smallsats and landers, bringing analysis to the data rather than shipping everything down. This is crucial for deep-space where bandwidth is thin.

Bottom line: "Think first, transmit second" saves power, time, and money.

11) Training the Team - Simulators That Learn

Humans still run the show. AI makes the training sharper.

- **Realistic sims:** ML-driven simulators model fluids, power systems, and communications glitches more realistically than hand-coded scripts. Crews and controllers practice rare failures they'll hopefully never see for real.
- **Adaptive tutoring:** Training software watches how a trainee solves problems and adjusts difficulty, pacing, and hints. Weak spots get extra reps before flight.
- **Procedure validation:** Language models stress-test checklists, searching for ambiguities, missing steps, or bad branching logic, before they ever reach orbit.

Bottom line: Practice makes permanent; AI makes practice better.

12) Limits, Risks, and the Rule of Good Sense

AI is a tool, not a wizard. A few grounded cautions:

- **Data bias:** Train on only "good days," and your model may fail on the bad one that matters. Include edge cases. Simulate ugliness. Seek diverse data.
- **Explainability:** A black-box model that screams "anomaly" without saying why is hard to trust in a crisis. Favor methods that show their work, feature importance, saliency maps, physical constraints.
- **Overfitting to noise:** Space environments change. A model "too perfect" on last year's data may fall apart next year. Regular retraining and sanity checks are mandatory.
- **Cybersecurity:** Anything connected can be attacked. AI must be designed with authentication, redundancy, and fail-safes that default to safe behavior.
- **Human in/on the loop:** For safety-critical calls, humans still own the "go/no-go." AI proposes; experienced people dispose.

Bottom line: Use AI like a power tool, strong, precise, and dangerous if you skip the guard.

13) Ethics and Policy - The Culture We Carry Into Orbit

Space exploration is a mirror. The choices we make in code show up in orbit.

- **Transparency:** Publishing methods and validation gives other teams confidence. Openness builds trust across agencies and countries.
- **Shared standards:** Common formats for alerts, health metrics, and risk scores make collaboration smoother in emergencies.
- **Debris responsibility:** AI that helps plan disposal and track fragments is part of being a good citizen in orbit.
- **Human dignity:** Crew monitoring must respect privacy while guarding health; data should be used with consent and clear purpose.

Bottom line: AI in space should make us **better neighbors**, not just better engineers.

14) Case Snapshots - Tiny Stories with Big Lessons

- **Autonomous landing:** A lander's camera sees a boulder field where the map predicted smooth ground. TRN compares the image to an onboard hazard map, shifts the target by a few hundred meters, and the lander touches down on safe terrain. Hours later the team reads the log and exhales.
- **Rover triage:** Overnight, a rover images a ridge. Onboard AI picks two frames with fine layering and flags them "priority." Those arrive first; geologists spot sedimentary clues and update the next drive target. A week saved, just like that.
- **Station maintenance:** A pump's vibration signature starts to drift. The model nudges the timeline: "swap before next eclipse

season." The crew replaces it during a calm afternoon. Boring success. The best kind.

- **Exoplanet hint:** A survey satellite finds a shallow, noisy dip. The model says "likely planet around a faint star." A ground telescope confirms. A junior student gets a first-author paper and a career.

Bottom line: The wins feel small in the moment. They compound.

15) What's Coming Next - The Frontier Isn't Just Out There

- **Foundation models for space data:** Large, pre-trained models that "speak" images, spectra, and time series from many missions at once. Fine-tune them for a new sensor with little data and get strong performance from day one.

- **Onboard language+vision copilots:** Ask your habitat: "Show me leaks over the past week and the likely source." It highlights three spots, the pressure log, and a recommended test. You still decide, but the prep is done.

- **Self-updating autonomy:** Spacecraft that learn new normal behaviors over long missions without forgetting their original training, while staying within strict safety bounds.

- **Swarms:** Dozens to hundreds of small satellites or surface bots that coordinate through distributed AI, mapping caves on the Moon, sniffing plumes on an icy moon, or building communications nets on Mars.

- **In-situ resource utilization (ISRU) robots:** Prospectors that identify water ice, mine regolith, sort grains, and bake bricks, guided by models that balance yield against wear and power.

- **Biofabrication and closed-loop life support:** AI steadies bioreactors and plant chambers, turning waste back into air, water, and food with fewer hiccups. The long voyage gets a reliable pantry.
- **Human–robot teams:** Crew lead, robots assist; robots lead, crew supervise, switching roles smoothly as tasks change from delicate science to heavy lifting.

Bottom line: We're moving from "AI helps missions" to "AI helps ecosystems of missions."

16) Tradition and Tomorrow - The Right Kind of Bold

You asked me to tell it straight and keep a steady, hopeful eye. Here it is:

- The tradition, **checklists, testing, calm voices, and speaking up about bad news**, is non-negotiable. Space will never forgive shortcuts.
- AI doesn't replace that tradition; it **extends** it. The same discipline that makes a great flight director makes a great AI pipeline: measure well, compare honestly, and fix what breaks.
- The future belongs to teams that **respect the craft** and **embrace the new tools**. Slide rules gave way to spreadsheets; blind scans gave way to trained models. The spirit is the same: **do the hard thing well**.

When we look up now, we're not just seeing a sky full of destinations; we're seeing a sky that **learns back**, that gives us patterns to read and choices to act on. AI is not magic. It's a sharp pencil in a moving hand.

Used wisely, it helps us land softer, see farther, waste less, and bring more of the universe home in our pockets.

That's the work. And it's good work.

Conclusion

We set out to read the sky like a ledger and found a poem written in numbers. The facts were firm - mass, thrust, orbits, budgets - and the feelings were earned: wonder without wishful thinking, courage without noise. Space isn't magic; it's craft. It rewards clean checklists, calm voices, and the habit of testing the hard parts twice. That's the old wisdom, and it still works.

What did we learn? That the universe is lawful and generous. That black holes, bright comets, quiet stations, and loud rockets all answer to the same rules. That cooperation beats swagger, and that "boring excellence" is how you get people home. We also learned that standing on Earth and looking up can make us better at caring for the ground under our boots.

So keep your tradition close and your tools sharp. Let curiosity power, but let discipline steer. Hand the wrench and the window seat to the next crew, and tell them what we were told: *measure twice, burn once; listen to bad news; test like you fly; wonder is earned by work.*

The sky isn't a ceiling. It's an invitation. Take it with steady hands, humble hearts, eyes wide.

Other Books by Buddy Johnston

The Beach – A Good Friend

An Intimate Reflection on a Familiar Place

The Beach – A Good Friend is a celebration of nature's most beloved playground. Through a series of heartfelt reflections, this book captures the joy, wonder, and timeless rituals that the shoreline offers. From the exhilaration of waves to the serenity of sandy sunsets, each page invites you to reconnect with the magic of the coast. Whether you're revisiting cherished memories or dreaming of your next seaside escape, this book reminds you to slow down, breathe deeply, and treasure the journey.

The Invisible Ledger

Stories of Karma Balancing the Books

The Invisible Ledger is a compelling collection of 50 real-life stories that prove one timeless truth: karma is alive, well, and not in the mood to be ignored. Through jaw-dropping twists, poetic justice, and soul-stirring moments, this book explores how the actions we take, whether good, bad, or somewhere in between, inevitably come back to meet us, often when we least expect it.

Drawing from everyday life and extraordinary experiences, each chapter unveils a different facet of karma at work: the bully brought low by his own schemes, the generous heart rewarded in the most unexpected way, the deceitful caught in their own tangled web. These aren't just cautionary tales. They are cosmic reminders that the energy we put out into the world doesn't vanish. It lingers. It circles back.

Grounded in ancient wisdom and sharpened by the edge of modern reality, this book shows that karma isn't superstition. It's universal law. It shaped the past, it governs the present, and it will echo far into the future, holding us all accountable not just to others, but to ourselves.

In a world where justice doesn't always show up in courtrooms or headlines, *The Invisible Ledger* reminds us that the universe has its own way of settling the score, and it never forgets.

We Dance!

A Reflective Look at a Dance Ministry for People Suffering from Dementia

This inspiring book takes you behind the scenes of Care to Dance, a nonprofit organization in central Florida that brings cognitive and emotional enrichment to residents in assisted living and memory care facilities through the art of dance. From the heartfelt stories of founders and volunteers to the moving testimonies of caregivers, activity directors and family members, you'll witness the profound impact of this unique program.

Most powerful of all are the moments when residents with dementia respond to the rhythm, movement and energy of the shows by smiling, clapping and even dancing along.

Let yourself be immersed in **We Dance!**, an uplifting journey that celebrates compassion, community and the simple yet powerful magic of dance.

A Collage of Magic

A Special Collection of All Things Magic

Step into a world of wonder, illusion and imagination.

In this captivating collection, Buddy Johnston explores the enchanting art of magic, its rich history, its evolving future and its vibrant present. Along the way, you'll meet legendary magicians and discover their signature illusions, gain insight into the craft, and enjoy stories from Buddy's personal experiences with magic.

At the heart of the book lies Milo Moon: The Boy Who Drew Lightning, a delightful fictional tale that captures the spirit of mystery and amazement that magic inspires in us all.

Whether you're a lifelong fan of illusion or newly curious, this book invites you to rediscover the joy of believing in the impossible. If you love magic, you'll love this book.

Cash Value Magic

Exploring the Benefits of Whole Life Insurance

Cash Value Magic invites readers into a quiet financial revolution where whole life insurance becomes more than a safety net. In just 35 crisp pages, Buddy Johnston reveals how a policy can evolve into a wealth-building engine that grows quietly, reliably, and purposefully over time.

This book demystifies the **cash value** component, a feature that accumulates savings within the policy while remaining accessible for life's evolving needs. Readers learn how to tap into this value to fund emergencies, retirement, or legacy goals, without disrupting long-term growth.

Johnston explores the magic of **Paid-Up Additions (PUAs)** and the compounding potential of **dividend reinvestment**, showing how these tools can expand both the cash value and death benefit with strategic elegance.

But this guide goes deeper than mechanics. It offers a vision for building wealth with soul, a financial path that honors loved ones, supports dreams, and echoes across generations.

With clarity, heart, and quiet confidence, *Cash Value Magic* empowers readers to rethink what insurance can be: not just protection, but possibility. For anyone seeking financial strategy with emotional resonance, this is the place to begin.

Five Fiery Romance Novellas

Tales that Trace the Lines Between Passion, Pain and Pleasure

Five Fiery Romance Novellas heats up when Savannah Blake and Julian Hart unexpectedly reunite at their friends' tropical destination wedding in ***Vows and Vendettas.*** Savannah and Julian are bitter enemies because of a relationship gone bad, but the wedding atmosphere causes them to re-evaluate their attitudes toward each other. Next, in ***Bitter River***, Juliet Vale, a stylish Manhattan powerbroker, and Eli Hawthorne, an ex-marine turned cowboy, are forced to spend 30 days together on a Montana ranch as stipulated by their fathers to earn their inheritances.

This is followed by ***Pier Pressure***, in which Councilwoman Trisha Spector is rescued by Jack Colton when she falls into the ocean after a portion of Paige's Pier collapses. Follow the twists and turns of this riveting story. Next, Executive Ethan Chase and muralist Lucia Reyes share time and experiences in Maimi and must make some tough decisions in ***The Confluence***. Finally, ***Northern Sparks*** features two mid-life entrepreneurs, Margot Delany and Graham Beckett snowed in at a convention in Banff, Alberta. This is a story of love after loss, creation after silence, and the kind of partnership that doesn't just ignite passion, it rewrites legacy.

Suspense Thriller Short Stories

A Collection of Riveting Tales to Keep You on the Edge of Your Seat

Read, if you dare, about a judge whose brother is a member of a drug cartel and whose son is a drug addict. In ***Family or Justice***, this judge has her loyalty put to the test. Then read about a working man who racked up large gambling debts, so he chose to kidnap the daughter of a wealthy businessman and demand a ransom to pay off his gambling debts. ***Desperate Odds*** has many twists and turns. Next, in ***Shadows in the Hollow***, you'll become familiar with a family who moved from St. Louis to a rural area in Tennessee to escape urban madness. What they got was not what they expected. In ***Karma Remembers*** you'll read about a private investigator, and former detective, who may have done some questionable things in his past, but he learns that karma has a special way of settling old scores. Finally, in ***The Third Silence***, another detective learns that he can't run from his past either, but he must face the circumstances so he can move on with his life.

Unbreakable

Inspirational Stories of Courage, Faith and the Human Spirit

Read about these courageous people who overcame enormous odds to accomplish great things. They rose above fear, pain and limitation and became symbols of courage for all of us. They stood up for what they believed and refused to back down. Their example is inspirational and a blueprint for living a successful life.

There is no quit in these people and they have the story to back it up. Read about how they touched the lives of those around them to achieve great things. This book is for anyone who has ever felt stuck, lost, overwhelmed or small. These stories are a reminder that inside everyone of us lives something unbreakable.

A Family of Romance Novellas

Love finds a thousand ways to be told. Five of them live here.

From the innocence of young sweethearts to the unexpected spark of strangers, these stories capture the timeless, unshakable power of love and determine destiny for a family of lovers.

In *A Rural Romance*, the true tale of David and Mamie Ruth shows how a high school devotion can ripple through generations. *The Savings Bond* offers the enduring pull of Jim and Kathy, bound together across a lifetime. *Round Trip No Expiration* reminds us how a single chance meeting can change everything for Mark and Susie. In *The Destiny Dance*, David and Debra discover that sometimes the rhythm of love is found on the dance floor. And *The No Shows* proves that even disappointment can lead to destiny, when Ron and Beth transform an abandoned evening into something unforgettable.

Five stories. Some true, some imagined. All united by one eternal truth: love endures unexpected, unshakable, and always extraordinary, Always.

From Logic to Learning to Liberation

From the gears of Charles Babbage's Analytical Engine to the neural networks powering today's generative models, this is the sweeping story of artificial intelligence, its birth, its winters, its dazzling rebirths, and the futures it may yet bring.

In these pages, you'll meet visionaries like Ada Lovelace, Alan Turing, and John McCarthy; follow the rise and fall of expert systems and neural networks; witness the triumphs of Deep Blue and AlphaGo; and step into today's era of AI copilots, transformers, and creative machines.

Told as a grand narrative, equal parts history, philosophy, and speculation, this book traces how humans taught machines to calculate, to reason, and to create, and asks the question echoing through every age: what does it mean to be intelligent? More than a chronicle of machines, this is a story about us, our ingenuity, our fears, our dreams, and the choices that will shape tomorrow.

www.ingramcontent.com/pod-product-compliance
Lightning Source LLC
LaVergne TN
LVHW051044210926
848486LV00001B/14